POLLUTION CONTROL IN THE PETROCHEMICALS INDUSTRY

M. Brett Borup
E. Joe Middlebrooks
Tennessee Technological University
Cookeville, Tennessee

LEWIS PUBLISHERS, INC.

Library of Congress Cataloging-in-Publication Data

Borup, M. Brett.
Pollution control in the petrochemicals industry.

Bibliography: p.
Includes index.
1. Petroleum chemicals industry – Waste disposal – Environmental aspects. 2. Pollution – Environmental aspects. I. Middlebrooks, E. Joe. II. Title.
TD899.P4B67 1987 661′.804 87-3724
ISBN 0-87371-039-8

LEWIS PUBLISHERS, INC.
121 South Main Street, P.O. Drawer 519, Chelsea, Michigan 48118

PRINTED IN THE UNITED STATES OF AMERICA

PREFACE

The petrochemicals industry is very complex and requires considerable knowledge of the individual processes to develop effective pollution control plans and processes. Information in this small book is intended to provide a base from which one can build. It is not exhaustive in describing the segments of the industry or pollution control techniques; however, it does provide a basic knowledge that should lead to intelligent, environmentally sound solutions to pollution prevention, control, and treatment.

The characteristics of various waste streams are summarized. A synopsis of treatment techniques and performance of these procedures is presented. All aspects of pollution control (air, water, solid wastes, hazardous wastes, and power consumption) are discussed.

Processes used to treat petrochemical manufacturing plants' wastes are as varied as the processes used in the manufacturing plants themselves. Pollution control costs vary significantly, and the basic approach to pollution control will significantly affect costs. Great care must be exercised when selecting pollution control equipment. Guidelines for the proper selection of pollution control equipment are presented throughout this book.

This book should meet the needs of many people: the engineer or scientist assigned the task of solving pollution problems; the consulting engineer seeking a solution for a client; the municipal engineer working with waste processing systems treating combined municipal and petrochemical wastes; the professor teaching an industrial waste processing course; and the state and federal regulatory agencies' personnel involved in the control of industrial pollution, the enforcement of standards, and the management of industries planning or operating waste processing facilities.

Many users of this book may find more detail than they require for a particular task, and others seeking more detail than is available may consult the reports and papers cited. Regardless of the needs of individuals, however, the book contains specifics or references that should make the task of pollution control easier.

We have made extensive use of many people's work and wish to thank these authors. Credit is given at the point in the text where the borrowed material appears.

We are especially grateful to Ms. Arlene Grover for her preparation of the original manuscript and the many hours of proofreading. Without her untiring efforts, the quality of the book would have suffered greatly. We wish to thank Messrs. Bob Hagemann and Ed Vail for the preparation of the final manuscript and Ms. Joann Carver for her typesetting efforts.

When people's contributions to such an extensive task as preparing a book are acknowledged, worthy candidates are unfailingly omitted. We apologize for any omissions and hope to do better next time.

CONTENTS

TABLES

FIGURES

POLLUTION CONTROL IN THE PETROCHEMICALS INDUSTRY

CHAPTER 1

INTRODUCTION TO THE PETROCHEMICAL INDUSTRY

The Industry

In 1980 the world consumption of 23 major petrochemical products was well over 250 million metric tons per year. World consumption of these products has been projected to grow at a rate of approximately 5 percent per year through the year 2000 (UNIDO, 1983). The size of the petrochemical industry, coupled with the nature and complexity of the wastes, produces the potential for significant impact on public health and environmental quality. The need for adequate waste management is obvious.

Petrochemicals, sometimes called petroleum-chemicals, can be defined as any chemicals which are derived from petroleum or natural gas. This definition includes a very wide variety of compounds from acetylene to vinyl chloride. Petrochemical compounds are included in several international standard industrial classifications including: industrial gases, cyclic intermediates, dyes, organic pigments and crudes, organic chemicals, inorganic chemicals derived from petroleum, plastic materials, insecticides and agricultural fertilizers.

Raw materials used in the production of petrochemicals include crude petroleum, natural gas, refinery gas, natural gas condensate, light tops or naphtha, and heavy fractions such as fuel oil. Primary or first generation petrochemicals are produced from these raw materials. Primary petrochemicals include alkynes such as acetylene; olefins such as ethylene, propylene, and butylene; paraffins such as methane, ethane, and propane; aromatics such as benzene, toluene, and xylene; hydrogen; hydrogen sulfide and carbon black. These first generation compounds are used as feedstocks in the synthesis of intermediate and third generation petrochemical products.

Primary, intermediate, and third generation compounds are produced by exposure of feedstocks to specific process conditions, which dictate the chemistry of the transformation. A wide variety of chemical reactions may be induced in these processes including: polymerization, hydration, halogenation, epoxidation, alkylation, hydrocarboxylation, nitration, sulfonation, oxidation, dehydrogenation and cracking, isomerization and crystallization. Processes are designed to favor the formation of some desired product; however, undesired compounds, which become waste products, are often formed.

Table 1-1. Petrochemical Processes as Waste Sources (Gloyna and Ford, 1970)

Process	Source	Pollutants
Alkylation: Ethylbenzene		Tar, Hydrochloric Acid, Caustic Soda, Fuel Oil
Ammonia Production	Demineralization	Acids, Bases
	Regeneration, Process Condensates	Amonia
	Furnace Effluents	Carbon Dioxide, Carbon Monoxide
Aromatics Recovery	Extract Water	Aromatic Hydrocarbons
	Solvent Purification	Solvents - Sulfur Dioxide, Diethylene Glycol
Catalytic Cracking	Catalyst Regeneration	Spent Catalyst, Catalyst Fines (Silica, Alumina) Hydrocarbons, Carbon Monoxide, Nitrogen Oxides
	Reactor Effluents and Condensates	Acids, Phenolic Compounds, Hydrogen Sulfide Soluble Hydrocarbons, Sulfur Oxides, Cyanides
Catalytic Reforming	Condensates	Catalyst (particularly Platinum and Molybdenum), Aromatic Hydrocarbons, Hydrogen Sulfide, Ammonia
Crude Processing	Crude Washing	Inorganic Salts, Oils, Water Soluble Hydrocarbons
	Primary Distillation	Hydrocarbons, Tars, Ammonia, Acids, Hydrogen Sulfide
Cyanide Production	Water Slops	Hydrogen Cyanide, Unreacted Soluble Hydrocarbons

Table 1-1. (continued)

Process	Source	Pollutants
Dehydrogenation		
Butadiene Prod. from n-Butane and Butylene	Quench Waters	Residue Gas, Tars, Oils, Soluble Hydrocarbons
Ketone Production	Distillation Slops	Hydrocarbon Polymers, Chlorinated Hydrocarbons, Glycerol, Sodium Chloride
Styrene from Ethyl-benzene	Catalysts	Spent Catalysts (Iron, Magnesium, Potasium, Copper, Chromium, and Zinc)
	Condensates from Spray Tower	Aromatic Hydrocarbons, including Styrene, Ethyl Benzene, and Toluene, Tars
Desulfurization		Hydrogen Sulfide, Mercaptans
Extraction and Purification		
Isobutylene	Acid and Caustic Wastes	Sulfuric Acid, C_4 Hydrocarbon, Caustic Soda
Butylene	Solvent and Caustic Wash	Acetone, Oils, C_4 Hydrocarbon, Caustic Soda, Sulfuric Acid
Styrene	Still Bottoms	Heavy Tars
Butadiene Absorption	Solvent	Cuprous Ammonium Acetate, C_4 Hydrocarbons, Oils
Extractive Distillation	Solvent	Furfural, C_4 Hydrocarbons
Halogenation (Principally Chlorination)		
Addition to Olefins	Separator	Spent Caustic

(continued)

Table 1-1. (continued)

Process	Source	Pollutants
Substitution	HCl Absorber, Scrubber	Chlorine, Hydrogen Chloride, Spent Caustic, Hydrocarbon Isomers and Chlorinated Products, Oils
	Dehydrohalogenation	Dilute Salt Solution
Hypochlorination	Hydrolysis	Calcium Chloride, Soluble Organics, Tars
Hydrochlorination	Surge Tank	Tars, Spent Catalyst, Alkyl Halides
Hydrocarboxylation (OXO Process)	Still Slops	Solule Hydrocarbons, Aldehydes
Hydrocyanation (for Acrylonitrile, Adipic Acid, etc.)	Process Effluents	Cyanides, Organic and Inorganic
Isomerization in General	Process Wastes	Hydrocarbons; Aliphatic, Aromatic, and Derivative Tars
Nitration		
Paraffins		By-Product Aldehydes, Ketones, Acids, Alcohols, Olefins, Carbon Dioxide
Aromatics		Sulfuric Acid, Nitric Acid, Aromatics
Oxidation		
Ethylene Oxide and Glycol Manufacture	Process Slops	Calcium Chloride, Spent Lime, Hydrocarbon Polymers, Ethylene Oxide, Clycols, Dichloride
Aldehydes, Alcohols, and Acids from Hydrocarbons	Process Slops	Acetone, Formaldehyde, Acetaldehyde, Methanol, Higher Alcohols, Organic Acids

Table 1-1. (continued)

Process	Source	Pollutants
Acids and Anhydrides from Aromatic Oxidation	Condensates Still Slops	Anhydrides, Aromatics, Acids Pitch
Phenol and Acetone from Aromatic Oxidation	Decanter	Formic Acid, Hydrocarbons
Carbon Black Manufacture	Cooling, Quenching	Carbon Black, Particulates, Dissolved Solids
Polymerization, Alkylation	Catalysts	Spent Acid Catalysts (Phosphoric Acid), Aluminum Chloride
Polymerization (Polyethylene)	Catalysts	Chromium, Nickel, Cobalt, Molybdenum
Butyl Rubber	Process Wastes	Scrap Butyl, Oil, Light Hydrocarbons
Copolymer Rubber	Process Wastes	Butadiene, Styrene Serum, Softener Sludge
Nylon 66	Process Wastes	Cyclohexane Oxidation Products, Succinic Acid, Adipic Acid, Glutaric Acid, Hexamethylene, Diamine, Adiponitrile, Acetone, Methyl Ethyl Ketone
Sulfation of Olefins		Alcohols, Polymerized Hydrocarbons, Sodium Sulfate, Ethers
Sulfonation of Aromatics	Caustic Wash	Spent Caustic

(continued)

Table 1-1. (continued)

Process	Source	Pollutants
Thermal Cracking for Olefin Production (including Fractionation and Purification)	Furnace Effluent and Caustic Treating	Acids, Hydrogen Sulfide, Mercaptans, Soluble Hydrocarbons, Polymerization Products, Spent Caustic, Phenolic Compounds, Residue Gases, Tars and Heavy Oils
Utilities	Boiler Blow-down	Phosphates, Lignins, Heat, Total Dissolved Solids, Tannins
	Cooling System Blow-down	Chromates, Phosphates, Algicides, Heat
	Water Treatment	Calcium and Magnesium Chlorides, Sulfates, Carbonates

Processes and Waste Streams

More than 500 different processing sequences are used in the petrochemical industry. The wide variety of process sequences coupled with the wide variety of products produced by the petrochemical industry leads to a complex waste problem. A list of the principal petrochemical processes and the wastes which may be expected to result from their use is presented in Table 1.1 (Gloyna and Ford, 1970). Examination of the information presented in this table reveals that many air and water pollutants along with solid wastes are generated during the production of petrochemicals.

Petrochemical wastes may produce a variety of adverse effects on public health and the environment. Biodegradable organic matter discharged to receiving waters may produce anaerobic conditions in the receiving water. These conditions will kill or drive off any aerobic organisms including fish and other higher animals. Anaerobic decomposition may also produce odor and color problems.

Thermal pollution from petrochemical discharges will also affect receiving waters, including death or decreased productivity of many aquatic species. Increased water temperatures also decrease oxygen solubility, enhance atmospheric oxygen transfer, and may produce increased biological activity. The net result will be a higher oxygen demand on the system.

Petrochemical plant discharges to receiving waters may also produce aesthetic effects such as objectionable odors, unsightly floating material, colored or turbid water, and foaming. These conditions may make a water unsuitable for recreational and other beneficial uses.

Wastestreams from petrochemical unit operations have also been found to contain toxic substances in many cases. Toxic substances associated with the production of some petrochemicals are listed in Table 1.2 (Wise and Fahrenthold, 1981). Several characteristics of these substances make them of particular concern. First, many of these substances are toxic at very low levels, sometimes in the ug/L range. Second, many of these compounds are biomagnified. This means high levels of these substances may be accumulated in organisms at high trophic levels. Third, many of these substances are refractory in nature. In other words, they are not easily degraded in the environment.

Petrochemical processing plants can also be significant sources of air pollution. A list of the air pollutants emitted from petrochemical processing plants and the major sources of these emissions is found in Table 1.3 (Mencher, 1967). Air pollution from petrochemical plants is produced by the combustion of fuel and by various losses from processing equipment.

Particulate matter, carbon monoxide, sulfur oxides and nitrogen oxides emissions are mainly a result of the combustion of fuels; however, other processes in the plant may cause these substances to be emitted. Hydrocarbon emissions may occur due to fuel combustion or various process losses, including leaking valves, flanges, pumps and compressors, evaporation from process drains, wastewater treatment processes, cooling water and blowdown systems, and losses from relief valves on operating and storage vessels. Research has shown that hydrocarbon emissions may be as great as 0.6%, by weight of total plant production. Hydrogen sulfide and mercaptans, produced during some production processes and emitted by various process losses, may create significant odor problems.

These air pollutants have been shown to have significant health effects. Air pollutants may adversely affect plant life, reducing crop yields, and plant growth

Table 1-2. Organic Chemical Effluents with Concentrations Greater than 0.5 ppm of Priority Pollutants (Wise and Fahrenthold, 1981)

Product	Generic process	Feedstock(s)	Associated priority pollutants
Acetone	Alkylation, peroxidation	Benzene, propylene	Aromatics
Acetylene	Dehydrogenation	Methane	Aromatics, polyaromatics
Acrolein	Oxidation	Propylene	Acrolein, aromatics, phenol
Acrylic acid	Oxidation	Propylene	Acrolein
Adiponitrile	Ammonolysis, dehydration	Adipic acid	Acrylonitrile
	Hydrodimerization	Acrylonitrile, hydrogen	Acrylonitrile
Alkyl (C13, C19) amines	Cyanation, hydrogenation	C12–C18 alpha olefin, HCN	Cyanide
Alkyl (C8, C9) phenols	Alkylation	Phenol, C8–C9 olefins	Phenol, aromatics
Allyl alcohol	Reduction (by alkoxide)	Acrolein, sec-Butanol	Acrolein, phenol, aromatics, polyaromatics
Aniline	Hydrogenation	Nitrobenzene	Aromatics
Benzene	Hydrodealkylation	Toluene	Aromatics, polyaromatics
	BTX extraction	Catalytic reformate	Aromatics

Table 1-2. (continued)

Product	Generic process	Feedstock(s)	Associated priority pollutants
	BTX extraction	Coal tar light oil	Aromatics, polyaromatics, phenols, cyanide
	BTX extraction	Pyrolysis gasoline	Aromatics, polyaromatics
Benzyl chloride	Chlorination	Toluene	Aromatics
Bisphenol A	Condensation	Phenol, acetone	Phenol, aromatics
Butadiene	Extractive distillation	C4 pyrolyzates	Acrylonitrile (acetonitrile solvent)
Butenes			Aromatics, polyaromatics
Butylbenzyl phthalate	Esterification	n-Butanol, benzyl chloride Phthalic anhydride	Phthalates
Caprolactam	Oxidation, oximation	Cyclohexane	Aromatics
	Dehydrogenation, oximation	Phenol	Aromatics, phenol
Carbon tetrachloride	Chlorination	Methane	Chloromethanes, chlorinated C2's
	Chlorination	Ethylene dichloride	Chloromethanes, chlorinated C2's

(continued)

Table 1-2. (continued)

Product	Generic process	Feedstock(s)	Associated priority pollutants
Chlorobenzenes	Chlorination	Benzene	Chloroaromatics, aromatics
Chloroform	Chlorination	Methane, methyl chloride	Chloromethanes, chlorinated C2's
m-Chloronitrobenzene	Chlorination	Nitrobenzene	Aromatics, nitro-aromatics, chloroaromatics
Creosote	Distillation	Coal tar light oil	Phenols, aromatics, polyaromatics
Cumene	Alkylation	Benzene	Aromatics
Cyclohexanol/-one	Oxidation	Cyclohexane	Phenol, aromatics
1,2-Dichloroethane	Oxychlorination	Ethylene, HCI	Chlorinated C2's
Dicyclopentadiene	Extraction, dimerization	C5 pyrolyzate	Aromatics, poly-aromatics
Diethylphthalate	Esterification	Ethanol, phthalic anhydride	Phthalates
Diketene	Dehydration	Acetic acid	Isophorone
Dimethyl terephthalate	Esterification	Methanol, TPA	Phthalates, phenol
Dinitrotoluenes	Nitration	Toluene	Nitroaromatics, aromatics, nitrophenols

Table 1-2. (continued)

Product	Generic process	Feedstock(s)	Associated priority pollutants
Diphenylisodecyl	Esterification	Phenol, isodecanol	Phenol, chlorophenols
phosphate ester		$POCI_3$	Aromatics
Epichlorohydrin	Chlorohydrination	Allyl chloride	Chlorinated C3's
Ethoxylates-alkylphenol	Ethoxylation	Alkylphenol, ethylene oxide	Phenol, aromatics
Ethylbenzene	Alkylation	Benzene, ethylene	Aromatics, poly-aromatics, phenol
	Extraction from BTX	BTX extract	Aromatics, poly-aromatics
			Acrylonitrile (acetronitrile solvent)
Ethylene	Steam pyrolysis	LPG, naphtha, or gas oil	Aromatics, poly-aromatics, phenol
Ethylene amines	Ammonation	1,2-Dichloroethane, NH_3	Chlorinated C2's
Ethylene diamine	Ammonation	1,2-Dichloroethane, NH_3	Chlorinated C2's
Ethylene oxide	Oxidation	Ethylene	1,2-Dichloroethane (Co_2 inhibitor)
Ethylene oxide	Chlorohydrination	Ethylene	Chlorinated C2's, chloroalkyl ethers

(continued)

Table 1-2. (continued)

Product	Generic process	Feedstock(s)	Associated priority pollutants
2-Ethylhexyl phthalate	Esterification	2-Ethylhexanol	Phthalates
		Phthalic anhydride	
Glycerine	Hydrolysis	Epichlorohydrin	Chlorinated C3's
Hexamethylene diamine	Hydrogenation	Adiponitrile	Acrylonitrile
Isobutylene	Extraction	C4 pyrolyzate	Aromatics
Isoprene	Extractive distillation	C5 pyrolyzate	Aromatics, polyaromatics
			Acrylonitrile (acetonitrile solvent)
Maleic anhydride	Oxidation	Benzene	Aromatics
Methacrylic acid	Cyanohydrination	Acetone	Cyanide
Methyl chloride	Chlorination	Methane	Chloromethanes, chlorinated C2's
	Hydrochlorination	Methanol	Chloromethanes
Methylene chloride	Chlorination	Methane	Chloromethanes, chlorinated C2's
		Methyl chloride	Chloromethanes, chlorinated C2's

Table 1-2. (continued)

Product	Generic process	Feedstock(s)	Associated priority pollutants
Methylethyl ketone	Reduction (alkoxide)	Acrolein, sec-Butanol	Acrolein, aromatics, polyaromatics,phenol
a-Methyl styrene	Peroxidation	Cumene	Aromatics, phenol
Naphthalene	Distillation	Coal tar distillates	Aromatics, poly-aromatics, phenols, cyanide
	Distillation	Pyrolysis gasoline	Aromatics, poly-aromatics
Nitrobenzene	Nitration	Benzene	Aromatics, nitro-aromatics
			Nitrophenols
Phenol	Peroxidation	Cumene	Aromatics, phenols
Phthalic anhydride	Oxidation	Naphthalene	Polyaromatics
	Oxidation	o-Xylene	Aromatics
Polymeric methylene dianiline	Condensation	Aniline, formaldehyde	Nitroaromatics
Polymeric methylene diphenyl diisocyanate	Phosgenation	Polymeric methylene dianiline, phosgene	Chloroaromatics (phosgenation solvent)
Propylene	Steam pyrolysis	LPG, naphtha, gas oil	Aromatics, poly-aromatics, phenols

(continued)

Table 1-2. (continued)

Product	Generic process	Feedstock(s)	Associated priority pollutants
Propylene oxide	Chlorohydrination	Propylene	chlorinated C3's, chloroalkyl ethers
Styrene	Dehydrogenation	Ethylbenzene	Aromatics, phenol
Tetrachloroethylene	Chlorination	1,2-Dichloroethane	Chloromethanes, chlorinated C2's
		RCI heavies	Chlorinated C3's
Tetrachlorophthalic anhydride	Chlorination	Phthalic anhydride	Chloroaromatics
Toluene	BTX extraction	Catalytic reformate	Aromatics
	BTX extraction	Coal tar light oil	Aromatics, poly-aromatics, phenols, cyanide
	BTX extraction	Pyrolysis gasoline	Aromatics
Tolylenediisocyanate	Phosgenation	Tolylenediamine	Chloroaromatics
1,2,4-Trichlorobenzene	Chlorination	1,4-Dichlorobenzene	Chloroaromatics
Trichloroethylene	Chlorination	1,2-Dichloroethane	Chlorinated C2's, chloromethanes
		RCI heavies	

Table 1-2. (continued)

Product	Generic process	Feedstock(s)	Associated priority pollutants
Vinyl acetate	Acetylation	Ethylene, acetic acid	Acrolein
Vinyl chloride	Dehydrochlorination	1,2-Dichloroethane	Chlorinated C2's, chloromethanes
Vinylidene chloride	Dehydrochlorination	1,1,2-Trichloroethane	Chlorinated C2's, chloromethanes
Xylenes (mixed)	BTX extraction	Pyrolysis gasoline	Aromatics
	BTX extraction	Catalytic reformate	Aromatics
	BTX extraction	Coal tar distillates	Phenols, aromatics, polyaromatics, cyanide

Table 1-3. Air Pollutants Emitted from Petrochemical Processing Plants (Mencher, 1967)

Pollutant	Source of pollutant
sulfur oxides	cracking units, treating units, flares, decoking operations, and all combustion operations
nitrogen oxides	combustion operations, compressor engines, catalyst regeneration
particulates	evaporation from storage tanks, loading facilities, sampling, spillage, processing equipment leakage, barometric condensors, cooling towers
carbon monoxide	combustion operations, decoking, catalyst regeneration
odors	hydrogen sulfide, mercaptans, wastewater treating units, barometric condensors

rate and in some cases causing the death of susceptible plants. Air pollution may also have corrosive effects on metals, building materials and textiles.

Management Philosophy

It is advantageous to consider excess materials as an additional resource to be utilized either in the form discarded or after further processing. This approach to waste processing is economically and environmentally important. If a government or ministry considers protection of the environment and maximum utilization of the base resource important, then the production management and the employees probably have an entirely different attitude toward performing this function and are more likely to take pride in producing high quality effluents and in recovering and utilizing as much of the material as possible. The importance of protecting the quality of the environment and the impact that improper handling of waste materials has on the employees' life styles and the nation as a whole must be emphasized (Middlebrooks, 1979).

Environmental protection must be stressed when management is expected to meet production quotas. Under such production systems management tends to concentrate its talent on product output, if not reminded continually of the value placed on environmental protection by the ministry and the nation. Environmental protection must be considered as a valuable natural resource in the same manner as the labor, materials, and the capital investment required to produce the basic product.

The costs for environmental protection must be paid either now or in the future. The most effective method of handling excess products is to incorporate the facilities for protecting the environment and for further processing of the excess into useful products. It is much less expensive to install such equipment initially than to convert a production process and add pollution control equipment later; moreover, it has proved cheaper to spend today's dollars than inflated ones of a later date. However, it is still less expensive to add to existing systems the facilities for processing excess materials than to allow excess to be wasted as environmental pollutants; to clean these up at a future time is costly and difficult. Indeed, the damage to the environment before installing equipment to correct a situation may be impossible to rectify. It is burdensome to assess the economic losses incurred by people and industry because of delayed pollution control; however, these are real economic factors which must be considered and emphasized. The losses of health, happiness, and productivity of people owing to environmental pollution are the greatest costs of all.

Long-term economic effects of industrial pollution must not be neglected. If an industry is allowed to develop in an area without pollution control facilities, eventually the area may deteriorate to a level unacceptable to many of the residents, and they move away. Relocation of the population depletes the tax base for public services and results in a further deterioration of the local living conditions. With an added tax burden the community is forced to extract more support from the industry, resulting in higher product costs. Environmental pollution also influences maintenance costs for homes, public buildings, and thoroughfares, as well as the industrial buildings and equipment themselves.

Pollution control is a good business practice which a nation cannot afford to neglect. Maintenance of the environment is much the same as maintenance of machinery, automobiles, and other devices: if a nation does not routinely care for the environment, eventually it deteriorates. In this case, deterioration may occur to a level that is intolerable to flora and fauna and cost the people and the government more than the industry produces. A nation must not sacrifice its customs and desirable environment to short-term economic advantage.

Summary

Some form of industrial waste treatment must be practiced if degradation of environmental quality is to be prevented. Complete treatment at the industrial site may be necessary, pretreatment prior to discharge to a public sewer may be required, or discharge to a treatment facility serving an industrial complex may provide the effluent quality needed. The degree of treatment required varies with local and national standards and the economy of byproduct recovery.

References

1. Gloyna, E. F. and D. L. Ford. 1970. The characteristics and pollutional problems associated with petrochemical wastes. Federal Water Pollution Control Administration, Research Series 12020.
2. Hedley, W. H., S. M. Menta, C. M. Moscowitz, R. B. Reznik, G. A. Richardson, and D. C. Zanders. 1975. *Potential Pollutants From Petrochemical Processes.* Technomic Publishing Co., Inc., Westport, Connecticut, USA.
3. Mencher, S. K. 1967. Minimizing waste in the petrochemical industry. *Chemical Engineering Progress,* 63(10):80.
4. Middlebrooks, E. J. 1979, *Industrial Pollution Control Vol I: Agro Industries.* John Wiley & Sons, New York, New York.
5. Wise, H. E., Jr. and P. D. Fahrenthold. 1981. Predicting priority pollutants from petrochemical processes. *Environ. Sci. and Techn.,* 15(11):1292.

CHAPTER 2

AIR POLLUTION

Air pollutants produced by petrochemical manufacturing practices include sulfur oxides, nitrogen oxides, carbon monoxide, particulates, odors, and a wide variety of hydrocarbons. These pollutants may be emitted from combustion operations (for energy and/or product production); cracking units, decoking operations and other unit processes; catalyst regeneration; flares; evaporation from storage tanks; spillage; leakage; cooling towers and condensers. The diversity and complexity of processes used in the petrochemical industry make it difficult to make sweeping generalizations about air pollutants emitted during petrochemical processing.

The U.S. EPA in an attempt to determine the significance of air pollution from the petrochemical industry conducted a study to determine industry descriptions, air emission control problems, sources of air emissions, statistics on quantities and types of emissions, and descriptions of emission control devices used. As a part of this survey, a method for rating the significance of air emissions was established and used to rank the processes studied and to select several processes for in-depth study.

The U.S. EPA studied a total of 33 distinctly different processes used to produce 27 petrochemicals, and these processes are listed in Table 2.1. The results obtained from this study are contained in a four volume series (Pervier et al. 1974 a, b, c, d). A summary of the estimated air emissions that would be emitted in 1980 from all of the plants utilizing these processes in the United States is shown in Table 2.2. The results shown in Table 2.2 are based on assorted sources of data and should be used as a guide to what might be expected but not as a rigorous comparison of process emissions. Some of the results are based on a 100 percent survey of the industry, while others are based on a limited data base or on engineering judgement.

In terms of mass of material emitted, the manufacture of carbon black leads by a large margin with the manufacture of acrylonitrile a distant second, emitting approximately one-third the mass of material discharged by carbon black. The mass of emissions per year is not the only criterion that must be considered in assessing the impact of an industry. The toxicity of emissions, odors and the persistence of the emitted compounds are some of the additional considerations. It can be seen from Table 2.2 that the petrochemical processing industry has the potential to be a significant source of air pollution.

Estimations of the air pollutants emitted, from 32 of the previously mentioned 33 processes, per unit weight of product produced may be found in Tables B.1 through B.32 in Appendix B (Pervier et al. 1974 a, b, c, d). These data were not published for

Table 2-1. Petrochemical Processes Surveyed (Pervier et al, 1974 a, b, c, d)

Acetaldehyde via Ethylene
Acetaldehyde via Ethanol
Acetic Acid via Methanol
Acetic Acid via Butane
Acetic Acid via Acetaldehyde
Acetic Anhydride
Adipic Acid
Adiponitrile via Butadiene
Adiponitrile via Adipic Acid
Polypropylene
Polystyrene
Polyvinyl Chloride
Styrene
Styrene-Butadiene Rubber
Vinyl Acetate via Acetylene
Vinyl Acetate via Ethylene
Vinyl Chloride via EDC Pyrolysis
Maleic Anhydride
Nylon 6
Nylon 6, 6
Oxo Process[a]
Phenol
High Density Polyethylene
Low Density Polyethylene
Carbon Disulfide
Cyclohexanone
Dimethyl Terephthalate (and Terephthalic Acid)
Ethylene
Ethylene Dichloride (Direct)
Formaldehyde (Silver Catalyst)
Glycerol (Allyl Chloride)
Hydrogen Cyanide (Andrussow)
Isocyanates via Amine Phosgenation

[a]Oxonation, or more properly, hydroformylation for the production of aldehydes and alcohols from olefins and synthesis gas.

the 33rd process, styrene. Estimations of hydrocarbon, particulate, sulfur oxides, nitrogen oxides, and carbon monoxide emissions are presented in these tables. The same precautions mentioned above should be exercised when using the data presented in Appendix B.

Information about specific pollutants emitted from petrochemical processing plants may be found in Hedley et al. (1975). One hundred and ninety petrochemical processes were identified, pollutant emissions from each process were identified, and emission stream compositions were tabulated (see Table 3.3 for a list of the 190 processes). Many of the individual hydrocarbons emitted from each of the processes are specifically named, and it is important to know specifics about discharges because of the wide range of toxicological properties of the substances.

Petrochemical plants discharge emissions into the atmosphere that are either

Table 2-2. Estimated Air Emissions from United States Petrochemical Plants (Pervier et al. 1974 a, b, c, d)

	ESTIMATED ADDITIONAL[a] AIR EMISSIONS IN 1980, MILLION LBS/YEAR					
	Hydrocarbons[b]	Particulates	Oxides of Nitrogen	Sulfur Oxides	Carbon Monoxide	Total
Acetaldehyde via Ethylene	1.2	0	0	0	0	1.2
via Ethanol	0	0	0	0	0	0
Acetic Acid via Methanol	0	0	0.04	0	0	0.04
via Butane	0	0	0	0	0	0
via Acetaldehyde	12.2	0	0	0	2.5	14.7
Acetic Anhydride via Acetic Acid	0.73	0	0	0	1.42	2.15
Acrylonitrile	284	0	8.5	0	304	596
Adipic Acid	0	0.14	19.3	0	0.09	19.5
Adiponitrile via Butadiene	10.5	4.4	47.5	0	0	62.4
Via Adipic Acid	0	0.5	0.04	0	0	0.54
Carbon Black	64	3.3	2.8	8.9	1,590	1,670
Carbon Disulfide	0.04	0.07	0.03	1.1	0	1.24
Cyclohexanone	77.2	0	0	0	85.1	162
Dimethyl Terephthalate (+TPA)	73.8	1.1	0.07	0.84	42.9	118.7
Ethylene	14.8	0.2	0.2	61.5	0.2	77
Ethylene Dichloride via Oxychlorination	110	0.5	0	0	25	136
via Direct Chlorination	34.2	0	0	0	0	34.2
Ethylene Oxide	32.8	0	0.15	0.05	0	33
Formaldehyde via Silver Catalyst	14.8	0	0	0	66.7	81.5
via Iron Oxide Catalyst	17.6	0	0	0	17.0	34.6
Glycerol via Epichlorohydrin	8.9	0	0	0	0	8.9
Hydrogen Cyanide Direct Process	0	0	0	0	0	0
Isocyanates	1.2	0.7	0	0.02	85	87
Maleic Anhydride	31	0	0	0	241	272
Nylon 6	0	3.2	0	0	0	3.2
Nylon 6, 6	0	5.3	0	0	0	5.3
Oxo Process	3.86	0.01	0.05	0	14.3	18.2
Phenol	21.3	0	0	0	0	21.3
Phthalic Anhydride via O-Xylene	0.3	13.2	0.8	6.8	113	134
via Naphthalene	0	0	0	0	0	0
High Density Polyethylene	210	6.2	0	0	0	216
Low Density Polyethylene	262	5	0	0	0	267
Polypropylene	152	0.5	0	0	0	152.5
Polystyrene	20	0.34	0	1.13	0	21.47
Polyvinyl Chloride	53	10	0	0	0	63
Styrene	3.1	0.05	0.1	0	0	3.25
Styrene-Butadiene Rubber	1.85	0.31	0	0.18	0	2.34
Vinyl Acetate via Acetylene	4.5	0	0	0	0	4.5
via Ethylene	0	0	TR	0	0	TR
Vinyl Chloride	26.3	0.9	0	0	0	27.2
Totals	1,547.2	55.9	79.5	80.5	2,588	4,351.9

[a] Assumes future plants will employ best current control techniques.
[b] Excludes methane, includes H_2S and all volatile organics.
[c] Includes non-volatile organics and inorganics.

controlled or fugitive in nature. Controlled emissions are released through stacks and/or vents, and detailed information is usually available on emissions composition and rate of release. Emissions from points other than stacks and vents are considered fugitive emissions. Fugitive emissions may occur due to accidents, inadequate maintenance, poor planning, and from a range of process equipment such as valves, pumps, flanges, compressors, and agitators.

Control of air pollutants emitted from controlled sources has been well studied. Many texts are available detailing the design of pollution control equipment for these sources (Crawford, 1976; Stern, 1968 a, b, and c; Danielson, 1967; Strauss, 1966; Strauss, 1972; Nonhebel, 1972; Lund, 1971; and Painter, 1974). A survey of the air pollution control techniques used in the petrochemical industry in India, along with gaseous pollutant regulations and monitoring techniques, may be found in an article by Ashar (1985). Due to the complexity of the processes used in petrochemical processing, it is difficult to make generalizations about the methods used to control air pollutant emissions at petrochemical plants; however, a few brief observations can be made.

The emission of carbon monoxide and particulate matter can be controlled with modern techniques of furnace design, proper fuel atomization and burner design. In other words, proper combustion process design will lead to lower pollutant emissions, while in most instances increasing the efficiency of the combustion process.

Oxides of nitrogen result from the high temperature combustion of fuels such as gas and oil. These nitrogen oxides, in the presence of hydrocarbons, and in sunlight, will produce photochemical smog with the conversion of nitric oxide to nitrogen dioxide (the more toxic nitrogen oxide) being accelerated. The amounts of nitrogen oxides emitted may be reduced by lowering the peak-flame temperatures of combustion processes, where the required reaction temperatures of the petrochemical processes will allow this temperature reduction.

Emissions of oxides of sulfur may be reduced in several ways. First, reduction of the sulfur content of the fuels used in the combustion process will result in lower sulfur oxides production. Desulfurization processes are available for coal, gas and liquid fuels (Economic Commission for Europe, 1984). Combustion processes may also be modified to produce lower sulfur emissions. Several processes are also available for the removal of sulfur compounds from combustion gases, including processes designed for recovery of sulfur compounds (lime/limestone process, sodium alkali process, dual alkali process and dilute sulfuric acid process), non-recovery processes (magnesium oxide process, sodium sulfite process, and aqueous carbonate process), dry removal processes, and combined removal of sulfur dioxide and oxides of nitrogen (Economic Commission of Europe, 1984).

Emissions of the fugitive type can occur from a range of circumstances and process equipment. Valves and flanges are the biggest contributors to fugitive losses at petrochemical plants. The large number of valves and flanges in a petrochemical processing plant means that even an average valve leak rate of 5 g/h will give a loss of 75 kg/h from valves alone at a typical olefin plant which may contain 15,000 valves. In fact, the U.S. EPA has determined that the average emission factor for valves at an olefin plant is 8.8 g/h (Jones, 1984). Based on this information it is not hard to justify efforts to reduce fugitive emissions from valves, flanges, and pumps since potential losses have been estimated to be over US $1800/day at a typical olefin plant.

Techniques used to measure fugitive emissions rates from petrochemical plants have been described by Hughes et al. (1979), Siversten (1983) and Ahlberg et al. (1985). Hughes et al. (1979) measured fugitive hydrocarbon emission rates at petrochemical plants manufacturing monochlorobenzene, butadiene, ethylene oxide/glycol and dimethyl terephthalate. Emission rates of the various sources measured at these plants are listed in Table 2.3.

Siversten (1983) conducted tracer experiments to quantify fugitive hydrocarbon emission rates at two petrochemical complexes. A simple proportionality model was applied to estimate leakage rates of ethylene, propylene, ethane, propane and isobutane from different parts of the complex. A dispersion model was then applied to verify concentration profiles and identify leakage areas.

A computer-automated CO_2 laser long-path absorption instrument was used by

Table 2-3. Hydrocarbon Emission Rates for Fugitive Sources by Process, g/hour (Hughes et al. 1979)

Source Type	Monochlorobenzene	Butadiene	Dimethylterephthalate	Ethylene Oxide/Glycol
	Average Emission Rate for Significant Fugitive Sources[a] (g/hr)			
Pump Seals	23	160	20	82
Compressor Seals	--	59	--	11
Valves	1.5	120	32	1.6
Flanges	82	0	110	1.0
Relief Devices	--	14	0	0
Process Drains	--	--	--	68
Agitator	200	--	218	--
Sample Valves	--	--	91	--
	Average Emission Rate for all Potential Sources[b]			
Pump Seals	7.7	63	3.3	13
Compressor Seals	--	54	--	5.9
Valves	0.05	17	1.5	0.07
Flanges	2.2	0	3.4	0.03
Relief Devices	--	5	0	0
Process Drains	--	--	--	40
Agitator	200	--	145	--
Sample Valves	--	--	40	--

Note: Dashes indicate source type nonexistent in process.

[a]Significant fugitive sources are those having an emission rate greater than or equal to 0.5 g/hr as determined by sampling and analysis.

[b]Emission rates were determined by calculating the mass of fugitive emissions from the emission rates for significant sources. The mass of emission was divided by the total number of sources screened to arrive at an average fugitive emission rate for all sources.

Ahlberg et al. (1985) to measure C_2H_4 emissions from a petrochemical plant. The laser system was used to estimate mass flow of C_2H_4 from point sources and fugative emissions sources. As laser and computer technology advance, techniques such as these may provide accurate measurements of fugative emissions.

Hydrocarbon emission reduction systems at petrochemical plants are described by Pruessner and Broz (1977), Kenson (1979), and Mashey and McGrath (1979). Pruessner and Broz (1977) described the design and operation of three incinerators, five condensation systems and two absorption systems for controlling hydrocarbon emissions. Tables 2.4 through 2.6 contain summaries of the operating conditions and costs of these systems. Treatment of waste air from three air oxidation processes by incineration achieved high removals (92-93%) of hydrocarbon contamination. Contaminant concentrations ranged from 150 ppm in a 1 million

Table 2-4. Operating Conditions and Costs of Incinerator Systems Used for Hydrocarbon Emissions Reduction (Pruessner and Broz, 1977).

	Maleic Incinerator	Oxo Incinerator	Houdry[a] "Puff" Reactor
Waste Gas Flow, lb/hr	220,000	235,000	900,000 (total) 13,000 (Puff reactor[b])
Contaminants, wt %			
Hydrocarbon	0.25	0.4	0.5
Carbon Monoxide	1.8	0.7	--
Removal Efficiency, %			
Hydrocarbon	93	93	92
Carbon Monoxide	95	95	--
Construction:			
Year	1975	1976	1975
Cost, US $	1,750,000	2,500,000	725,000
Heat Efficiency, %	85	82	80
Natural Gas Added, Std. cu ft/hr	80,000	130,000	0
Retention Time, Sec	0.7	0.5	0.3

[a]Modification of Houdry butane dehydrogenation process where hydrocarbon pollutants are concentrated in about 1% of the reheat air flow, thereby significantly reducing the discharge of pollutants.

[b]Waste gas flow diverted through the "Puff" reactor. This waste stream contains most of the hydrocarbon pollutants.

lb/hr gas flow rate to 4,000 ppm in a 235,000 lb/hr gas flow rate. Large volumes of natural gas are required to treat these large flows at high temperatures; therefore, energy recovery is an important part of these systems (Table 2.4). Condensation was used to remove up to 98 percent of a high hydrocarbon concentration from small noncondensable waste gas flows. The main advantages of condensation are product recovery and a relatively low energy requirement to remove the pollutants (Table 2.5). The two absorption systems utilized oil to absorb hydrocarbon from nitrogen waste gas streams. The oil is a mixture of parafinnic and aromatic oils which reduces the tendency for polymerization of the hydrocarbons contained within the tower. Fresh oil starts at the top of a five-stage tower and progresses through each stage, and at the bottom the oil contains about 3% hydrocarbon. Overall removal efficiency for hydrocarbons exceeds 98 percent for a gas flow rate of 275 lb/hr

Table 2-5. Operating Conditions and Costs of Condensation Systems Used for Hydrocarbon Emissions Reduction (Pruessner and Broz, 1977)

	Neoprene Monomer Isomerization Tower	Neoprene Monomer[a] Topping Column	Neoprene Polymer[a] Vessel Vents	Neoprene Latex Stripper Vent	Neoprene Polymer Emergency Dump System	
Type of Heat Exchanger	S&T[b]	S&T	DC[c]	S&T	DC	
Waste Gas Flow, lbs/hr Hydrocarbon	159	---	126	1,140	15,200	
Waste Gas Flow, lb/hr Total	331	542	275	2,875	32,000	
Hydrocarbon Removal Efficiency, %	81	99	43	99.8	99.995	
Heat Load, Btu/hr	22,000	93,000	110,000	1.2Mil	10,000	Steady State 3 Million Heat Sink/Dump
Operating Temperature, °F	-2	-2	.36	-2	40 to 75	
Construction:						
Year	1973	1973	1974	1969	1974	
Cost US $	20,000	30,000	40,000	120,000	250,000	

[a] Waste gas exiting this system is further treated in absorption system.
[b] Shell-and-Tube.
[c] Direct contact with water

Table 2-6. Operating Conditions and Costs of Absorption Systems Used for Hydrocarbon Emissions Reduction (Pruessner and Broz. 1977)

	Neoprene Monomer Absorber	Neoprene Polymer Vent Absorber
Spray Tower Stages	2	5
Waste Gas Flow to ABS[a] Hydrocarbon, lbs/hr	31	72
Waste Gas Flow, Total, lb/hr	36	187
Absorber Efficiency, %	90	97
Heat Load, Btu/hr	13,000	330,000[a]
Operating Temperature, °F	65	45
System Efficiency Including Condensation, %	99.5	98.4
Construction		
Year	1975	1974
Cost US $	60,000	300,000

[a] Includes heat load for recovery of hydrocarbons

containing 125 lb/hr hydrocarbon contamination. The recovered hydrocarbon is stripped from the oil and used in the process (Table 2.6).

Operating costs were not discussed for these systems. In the case of the condensation and absorption systems, hydrocarbons removed from the waste gas flow were returned to the process for further utilization. This would lower the net operating costs of the processes significantly, possibly resulting in a benefit rather than a cost.

Kenson (1979) reported on the engineered design of systems for organic emissions control at petrochemical plants. Two examples of toxic emission control were discussed, vinyl chloride monomer emission control at a polyvinyl chloride plant and benzene emission control. Multiple radial carbon beds in a single tank with regeneration by vacuum and indirect steam heating removed greater than 99 percent of the vinyl chloride monomer (VCM) emissions. By using vacuum and indirect carbon bed heating, the steam condensate was not contaminated with VCM. The carbon beds are capable of producing an effluent containing less than 5 ppm of VCM, and large quantities of VCM can be recovered for reuse in the PVC manufacturing process. An economic analysis showed that the value of the VCM recovered can pay for the VCM control system in three years.An economic analysis of a carbon adsorption system designed to control benzene emissions showed a credit of US $43,000 per year as a result of solvent recovery (Kenson, 1979). The following five concepts were presented for the design of an organic chemical emission control system:

1. The problem to be solved must be defined as thoroughly as possible. This requires a careful analysis of the temperature composition and volume flow rate of the exhaust stream, including the maximum and minimum values. Particulate concentration/size data may also be required.
2. The degree of control required must be well defined. This will allow the proper evaluation and selection of all the alternative control systems which might achieve this control efficiency.
3. The technical advantages and disadvantages of all the alternate control systems capable of achieving the desired degree of control must be weighed before final selection. If this is not done before system choice is made, the best control concept for that particular application may be passed up.
4. The total cost (capital and operating) of the alternative control system, including energy consumption and energy price sensitivity, must be evaluated to find which is most cost-effective for that application. Otherwise, the choice may be for the lowest capital cost system, which may be exceeded in 1-2 years by the cost of energy consumption in that system.
5. The final system choice must be designed to optimally control that particular exhaust stream. If a standard off-the-shelf system is used, it may give less than the desired degree of control and may have excessive operating costs. An engineered system may cost no more than an off-the-shelf solution.

Mashey and McGrath (1979) described another approach to the engineered design of organics emission control systems in which a detailed explanation is given of the design of the vapor collection systems necessary to transport emissions to control devices. Various types of control systems were discussed including thermal oxidation systems, catalytic oxidation systems, carbon adsorption, and gas compression/condensation systems. Cantrell (1982) also described techniques of organic vapor recovery at petrochemical plants.

Some other techniques used to reduce hydrocarbon emissions include: 1) appropriate specification, selection , and maintenance of seals in valves, pumps, and flanges (Jones, 1984), 2) installation of floating roof tanks to control evaporation of light hydrocarbons, 3) installation of vapor recovery lines to vents of vessels that are continually filled and emptied, 4) manifolding of purge lines used for start-ups and shutdowns to vapor recovery or flare systems, 5) venting of vacuum jet exhaust lines to vapor recovery systems, 6) shipment of products by pipeline rather than car or truck, 7) covering of wastewater separators, and 8) the use of steam or air injection at flares (Mencher, 1967).

Excellent case studies which include data on plant emissions, control devices, and cost effectiveness may be found in a report prepared by Air Products and Chemicals, Inc. (1974 a, b; 1975 a, b, c, d, e, f, and g). These reports contain detailed information about nine of the industries surveyed by Pervier et al. (1974 a, b, c, and d), mentioned previously. Because of the diversity of the unit operations involved in this industry and the complexity of the air emissions control problem, space does not

Table 2-7. Cost Effectiveness of Alternative Emission Control Devices in Polyvinyl Chloride Manufacture. (Based on 200 million lbs/yr PVC Plant) (Air Products and Chemicals, Inc. 1975 g)

	Plant W/O Control	Model Plant I	Model Plant II
Vinyl chloride monomer (VCM) Emissions, kg/kg of product			
Source			
A. Solution Storage	0.0030	0.0015	0.0005
B. Precipitation Tank	0.0030	0.0020	0.0005
C. Slurry Tank	0.0032	0.0010	0.0010
D. Crude Solvent Storage	0.0048	0.0015	0.0003
E. Blend Tank	0.0042	0.0015	0.0005
F. Centrifuge	0.0013	0.0002	0.0002
G. Bin Storage	0.0070	0.0030	0.0010
H. Fugitive	0.0080	0.0050	0.0030
Total	0.0345	0.0157	0.0070
Capital Cost of Control Devices			
High vacuum and compressor for maximum stripping		750,000	750,000
Refrigeration on condenser		70,000	
Substitute canned pumps		10,000	10,000
Monitoring equipment		175,000	175,000
Scrubber for VCM Recovery Vent System			125,000
Gas Holder			950,000
Solvent Cleaning of Reactors			300,000
		$ 1,005,000	$ 2,310,000
Operating Cost			
Cooling Water required, gpm		35	115
Electric power required, kwh/hr		95	126
Labor, men/shift		2.5	3
Steam, lbs/hr			1,000
Chemicals, US $/yr			62,100

permit a detailed case history of the plants. Summaries of efficiencies and economics of control devices which may be used to control pollution produced in the production of polyvinyl chloride and vinyl chloride monomer are presented in Tables 2.7 and 2.8. Anaylsis of the data contained in these tables illustrates the relationship between relative pollution control effectiveness and cost. In Table 2.7, the data show that 54 percent of the pollutants were removed at a capital investment of US $1,005,000. The removal of an additional 26 percent of the hydrocarbons cost approximately 2.3 times as much, or US $2,310,000, and the operating costs

Table 2-8. Cost Effectiveness of Alternative Emission Control Devices in Vinyl Chloride Monomer Manufacture by the Balanced Process. (Based on 200 million lbs/yr PVC Plant) (Air Products and Chemicals, Inc. 1975 f)

	Existing Plant	Model Plant I	Model Plant II
Vinyl chloride monomer (VCM) Emissions, kg/kg of product			
Ethylene dichloride (EDC) Distillation Column	0.000500	0.000100	0.000100
Scrubber Vent Stack	0.002400	0.000600	0.000048
Loading Losses	0.000796	0.000119	0.000119
Sampling	0.000038	0.000009	0.000009
Neutralizers and Filters	0.000003	0.000003	0.000003
Process Vessels	0.000078	0.000078	0.000078
Oxychlorination Vent	0.001320	0.001320	0.000026
Fugitive	0.000300	0.000225	0.000150
Total	0.005435	0.002454	0.000533
Capital Cost of Control Devices (US $)			
Refrigeration		200,000	--
Waste Heat Boiler		--	300,000
Incineration & Water Heat Boiler		--	1,140,000
Compressor and Refrigeration		200,000	200,000
Continuous Loop Sampler		50,000	50,000
Canned Pumps		--	200,000
Monitoring VCM Leaks		200,000	200,000
Total		$ 650,000	$ 2,090,000
Operating Costs			
Electricity, kwh/hr		80	185
Cooling water, gpm		60	30
Process water, gpm			90
Boiler feed water, gpm			82
Caustic lb/hr			1,350
Fuel million BTU/hr			16-30
Steam generated lbs/hr			38,800

increased significantly. This same trend is seen in Table 2.8 where a 55 percent pollutant removal is accomplished with a capital investment of US $650,000. An additional 35 percent of the pollution is removed at 3.2 times the cost, or US $2,090,000. Again, the operating costs increased significantly. These data show that the higher the removal rate, the higher the removal cost. For this reason, the selection of emission standards which must be met is very important.

As noted previously, many techniques which reduce air emissions produce

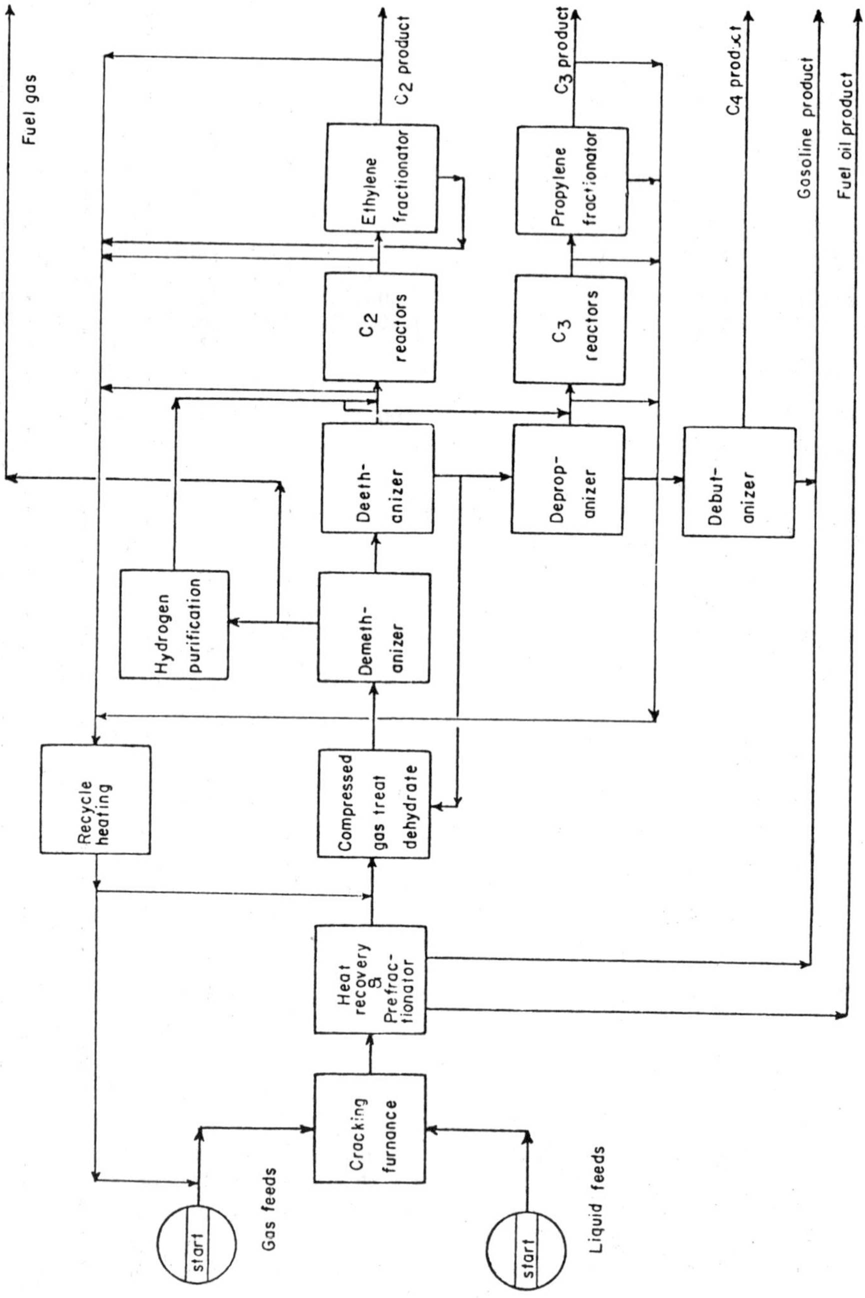

Figure 2.1. Simplified Flow Scheme for Ethylene Plant Showing Recycle of Off-Specification Products to Reduce Air Emissions and Produce Economic Savings. (Max and Jones, 1983).

economic benefits by reducing product loss and recovering usable compounds. Max and Jones (1983) reported on an operation technique that not only reduced air emissions but reduced production costs at an ethylene plant. Off specification products produced during start-up, shutdown, and upsets are recycled through the process train. The authors reported that as much as US $250,000 can be saved per start-up including product and feedstock losses. Figure 2.1 is a simplified flow scheme for this technique.

References

1. Ahlberg, H., S. Lundqvist and B. Olsson. 1985. Carbon dioxide laser long-path measurements of diffuse leakages from a petrochemical plant. *Applied Optics*, 24(22): 3924-3928.
2. Air Products and Chemicals, Inc. 1974a. Engineering and cost study of air pollution control for the petrochemical industry. Volume I: carbon black manufacture by the furnace process. U.S. EPA, EPA-450/3-73-006-a. Research Triangle Park, NC. NTIS#PB-238 324.
3. Air Products and Chemicals, Inc. 1974b. Engineering and cost study of air pollution control for the petrochemical industry. Volume III: ethylene dichloride manufacture by oxychlorination. U.S. EPA, EPA-450/3-73-006-c. NTIS#PB-240 492.
4. Air Products and Chemicals, Inc. 1975a. Engineering and cost study of air pollution control for the petrochemical industry. Volume II: Acrylonitrile Manufacture. U.S. EPA, EPA-450/3-73-006-b. NTIS#PB-240 986.
5. Air Products and Chemicals, Inc. 1975b. Engineering and cost study of air pollution control for the petrochemical industry. Volume IV: formaldehyde manufacture with the silver catalyst process. U.S. EPA, EPA-450/3-73-006-d. NTIS#PB-242 118.
6. Air Products and Chemicals, Inc. 1975c. Engineering and cost study of air pollution control for the petrochemical industry. Volume V: formaldehyde manufacture with the mixed oxide catalyst process. U.S. EPA, EPA-450/3-73-006-e. NTIS#PB-242 547.
7. Air Products and Chemicals, Inc. 1975d. Engineering and cost study of air pollution control for the petrochemical industry. Volume VI: ethylene oxide manufacture by direct oxidation of ethylene. U.S. EPA, EPA-450/3-73-006-f. NTIS#PB-244 116.
8. Air Products and Chemicals, Inc. 1975e. Engineering and cost study of air pollution control for the petrochemical industry. Volume VII: phthalic anhydride manufacture from ortho-xylene. U.S. EPA, EPA-450/3-73-006-g. NTIS#PB-245 277.
9. Air Products and Chemicals, Inc. 1975f. Engineering and cost study of air pollution control for the petrochemical industry. Volume VIII: vinyl chloride manufacture by the balanced process. U.S. EPA, EPA-450/3-73-006-h. NTIS#PB-242 247.

10. Air Products and Chemicals, Inc. 1975g. Engineering and cost study of air pollution control for the petrochemical industry. Volume IX: polyvinyl chloride manufacture. U.S. EPA, EPA-450/3-73-006-i. NTIS#PB-247 705.
11. Ashar, N. G. 1985. Control of gaseous pollutants in industrial emissions. *Indian Journal of Environmental Protection.* 5(2): 109-119.
12. Cantrell, C. J. 1982. Vapor Recovery for Refineries and Petrochemical Plants. *Chemical Engineering Progress,* 78(10):56-60.
13. Crawford, M. 1976. *Air Pollution Control Theory.* McGraw-Hill, Inc., New York, NY.
14. Danielson, J. A., ed. 1967. *Air Pollution Engineering Manual,* Public Health Service Publication 999-AP-40, National Center for Air Pollution Control, Cincinnati, OH (available from U.S. Government Printing Office, Washington, DC).
15. Economic Commission for Europe. 1984. *Air-borne Sulphur Pollution.* United Nations, New York.
16. Hedley, W. H., S. M. Mehta, C. M. Moscowitz, R. B. Reznik, G. A. Richardson, D. L. Zanders. 1975. *Potential Pollutants from Petrochemical Processes.* Technomic Publications, Westport, CT.
17. Hughes, T. W., D. R. Tierney, and Z. S. Khan. 1979. Measuring fugitive emissions from petrochemical plants. *Chemical Engineering Progress,* 75(8):35-39.
18. Jones, A. L. 1984. Fugitive emissions of volatile hydrocarbons. *Chemical Engineer* (London); Volume 406, Aug.-Sept. 1984, pp. 12-13; 15.
19. Kenson, R. E. 1979. Engineered Systems for the Control of Toxic Chemical Emissions. Proceedings of the 72nd Annual Meeting of the Air Pollution Control Association. Paper #79-17.4.
20. Lund, H. F., ed.1971. *Industrial Pollution Control Handbook.* McGraw-Hill Book Company, New York, NY.
21. Mashey, J. H. and J. J. McGrath. 1979. Engineering Design of Organics Emission Control Systems. Presented at the 1979 National Petroleum Refiners Association Annual Meeting.
22. Max, D. A. and S. T. Jones. 1983. Flareless ethylene plant. *Hydrocarbon Processing* 62(12):89-90.
23. Mencher, S. K. 1967. Minimizing Waste in the Petrochemical Industry. *Chemical Engineering Progress,* 63(10):80-88.
24. Nonhebel, G., ed. 1972. *Processes for Air Pollution Control,* CRC Press, Cleveland, OH.
25. Painter, D. E. 1974. *Air Pollution Technology.* Reston Publishing Company, Inc., Reston, VA.
26. Pervier, J. W., R. C. Barley, D. E. Field, B. M. Friedman, R. B. Morris, W. A. Schwartz. 1974a. Survey Reports on Atmospheric Emissions from the Petrochemical Industry, Volume I. U.S. EPA, Research Triangle Park, N.C. EPA-450/3-73-005-a.
27. Pervier, J. W., R. C. Barley, D. E. Field, B. M. Friedman, R. B. Morris, W. A. Schwartz. 1974b. Survey Reports on Atmospheric Emissions from the Petrochemical Industry, Volume II. U.S. EPA, Research Triangle Park, N.C. EPA-450/3-73-005-b.

28. Pervier, J. W., R. C. Barley, D. E. Field, B. M. Friedman, R. B. Morris W. A. Schwartz. 1974c. Survey Reports on Atmospheric Emissions from the Petrochemical Industry, Volume III. U.S. EPA, Research Triangle Park, N.C. EPA-450/3-73-005-c.
29. Pervier, J. W., R. C. Barley, D. E. Field, B. M. Friedman, R. B. Morris, W. A. Schwartz. 1974d. Survey Reports on Atmospheric Emissions from the Petrochemical Industry, Volume IV. U.S. EPA, Research Triangle Park, N.C. EPA-450/3-73-005-d.
30. Pruessner, R. D. and L. B. Broz. 1977. Hydrocarbon Emission Reduction Systems. *Chemical Engineering Progress*, 73(8):69-73.
31. Siversten, B. 1983. Estimation of diffuse hydrocarbon leakages from petrochemical factories. *Journal of the Air Pollution Control Association* 33(4):323-327.
32. Stern, A.C., ed. 1968a. *Air Pollution* Volume I. "Air Pollution and its effects". Academic Press, Inc., New York, NY.
33. Stern, A.C., ed. 1968b. *Air Pollution* Volume II. "Analysis, Monitoring and Surveying". Academic Press, Inc., New York, NY.
34. Stern, A. C., ed. 1968c. *Air Pollution* Volume III. "Sources of Air Pollution and their Control". Academic Press, Inc., New York, NY.
35. Strauss, W. 1966. *Industrial Gas Cleaning.* Pergamon Press, London, England.
36. Strauss, W. 1972. Air Pollution Control, Parts I and II. Interscience Publishers, New York, NY.

CHAPTER 3

WATER POLLUTION

Wastewater Streams

Wastewater streams in the petrochemical production industry may be categorized into six logical source components (Ford and Tischler, 1974):

(1) wastes discharged directly from production units during normal operation
(2) utility operations such as blowdown from energy production and cooling systems
(3) sanitary sewage from administrative areas, locker rooms, shower and restroom facilities, and food handling areas
(4) contaminated storm runoff from process areas
(5) ballast water discharged from tankers during product handling
(6) miscellaneous discharges from spills, turnarounds, etc.

The many combinations of production processes make it difficult to make generalizations about petrochemical wastewaters; however, petrochemical wastes may include various chemicals derived from petroleum derivatives and natural gas, toxic substances, lubricants, gas oil, fuel oil, wax, asphalt and petroleum coke. The hydrocarbons found in these wastestreams generally originate from leaks, spills, and product dumps. Steam condensate from reflux systems may contain hydrogen sulfide and mercaptans. Caustics, when used to purify hydrocarbon streams, produce alkaline wastestreams which are potentially toxic.

Ammonia may be introduced into petrochemical wastestreams from two sources: it may be added to product streams for corrosion control, and by the breakdown of nitrogenous compounds present in the feedstock. Other components of petrochemical wastestreams which may be of concern are corrosion inhibitors, particularly heavy metals.

Wastewater Characteristics

Gloyna and Ford (1970) conducted a survey designed to characterize petrochemical production wastes and to define the pollution problems associated with these wastes. Effects were described of petrochemical wastewater streams on receiving waters, on water used for other beneficial uses, and in-plant reuse. Several petrochemical wastewaters were also described in terms of conventional pollutional parameters such as acidity, alkalinity, color and turbidity, pH, biochemical oxygen demand (BOD), chemical oxygen demand (COD), total organic carbon (TOC), solids, surface activity, taste and odor and temperature. The results of effluent analyses from several typical petrochemical plants are presented in Table 3.1 (Gloyna and Ford, 1970).

The data shown in Table 3.1 illustrate the variability of waste characteristics in the petrochemical industry. The pH values of petrochemical wastewaters are generally greater than 7, and the wastestreams typically contain large amounts of total solids and low concentrations of suspended solids, indicating that most solids in these wastewaters are in the dissolved form. The variability in the data found in Table 3.1 suggests that each petrochemical wastestream must be analyzed separately to predict its characteristics. The variability can be attributed to the large number of choices of processes that may be selected to form a petrochemical plant.

Table 3-1. Total Plant Effluent Analyses, Typical Petrochemical Plants (Process Waste Before Treatment) (Gloyna and Ford, 1970)

Plant Products	Mixed Chemicals incl. ethylene oxide, propylene oxide, glycols, amines, and ethers	Refinery, Detergent Alkylate	Refinery Butadiene, Butyl Rubber
Alkalinity (mg/L)	4,060	365	164
BOD (mg/L)	1,950	345	225
Chlorides (mg/L)	430-800	1,980	825
COD (mg/L)	7,970-8,540	855	610
Oils (mg/L)	547	73	-
pH	9.4-9.8	9.2	7.5
Phenols (mg/L)	-	160	17
Sulfates (mg/L)	655	280	-
Suspended Solids (mg/L)	27-60	121	110
TOC (mg/L)	-	-	160**
Total Nitrogen (mg/L)	1,160-1,253	89	48
Total Solids (mg/L)	2,191-3,029	3,770	2,810
Misc. as Indicated		Sulfide= 150 ppm*	PO_4=trace

* Cooling Water Excluded

** Filtered

Table 3-1. (continued)

Plant Products	Mixed Organics	2,4,5-Tri chloro-phenol	2,4-Dichloro-phenol	Nylon
BOD_5 (mg/L)	1,950	16,800	16,700	170
Chlorides (mg/L)	800	96,300	144,000	800
COD (mg/L)	1,972	21,700	27,500	2,000
Oils (mg/L)	547	---	---	45
Phenols (mg/L)	10-50	---	---	400
Suspended Solids (mg/L)	60	700	348	neg.
Total Nitrogen (mg/L)	1,253	40	45	100
Total Solids (mg/L)	3,029	172,467	167,221	3,000
Misc. as Indicated	SO_4=655 mg/L	%Vol.TS= 10.5	%Vol.TS=13.2	H_2S=12 mg/L

The most commonly used method for predicting the quality and quantity of petrochemical production wastewaters is to study each individual unit process and relate the quantity and quality of the wastestreams produced to production units. For example, the isopropanol stripping still and intermediate flash column used in acetone production produces approximately 2.2 pounds of COD per ton of acetone produced (Hedley et al. 1975). The nature of production and processing can make this a difficult task. Small changes in unit process operating conditions such as temperature, pressure, flowrate, and variations in feedstock quality may significantly alter the characteristics of the wastestreams produced. A knowledge of the chemistry involved in the process, process operating conditions, feedstock used, and quantity of product produced by the unit operation can lead to the estimation of pollutant characterization and quantification.

The BOD and COD of many organic compounds that may be produced during the production of petrochemical products is available from several sources. Bridie et al. (1979) has presented the COD, BOD and theoretical oxygen demand for 118 petrochemical compounds. Other BOD and COD data may be found in Pritter (1976), Price et al. (1974), Meinck et al. (1968), and Heukelekian and Rand (1955).

Many toxic substances may be produced during the production of petrochemical products. The U.S. Environmental Protection Agency (US EPA) has identified 129 toxic organic chemicals which have been found in the waters of the nation, and these

Table 3-1. (continued)

Plant Products	Phenols, Cresols
Alkalinity (mg/L)	192
BOD_5 (mg/L)	550–850
Chlorides (mg/L)	230
COD (mg/L)	990–1,940
Color (Color Units)	50
Hardness (mg/L)	250
IOD (mg/L)	17
Kjeldahl-N (mg/L)	trace
NH_3-N (mg/L)	trace
Oil (mg/L)	trace
pH	4.6–7.2
Phenols (mg/L)	280–550
PO_4 (mg/L)	3
Sulfides (mg/L)	trace–1
Suspended Solids (mg/L)	12–88
Temperature ^oC	24.5
TOC (mg/L)	320–580
Total Solids (mg/L)	1,870–2,315

chemicals have become known as "priority pollutants." Wise and Fahrenthold (1981) presented a method which can be used to predict the occurrence of these 129 "priority pollutants" in petrochemical processing wastewaters. Critical precursor and generic process combinations that generate "priority pollutants" in 172 petrochemical manufacturing effluents are reported in Table 3.2 (Wise and Fahrenthold, 1981).

Hedley et al. (1975) conducted a survey of the processes associated with petrochemical production designed to characterize the wastes produced. The major potential sources of pollutants from the production of petrochemicals were identified. Published and unpublished data were used to describe emission sources and the composition of the emission streams from the 190 petrochemical production processes studied (Table 3.3). Hedley et al. (1975) presented process characterization sheets for each of these 190 processes which included a brief description of the process, utility requirements, feed materials, emissions sources, and potential pollutants. The complexity of the industry is apparent from the list of processes shown in Table 3.3. If details are needed for a specific process the report should be consulted.

Treatment Methods

The design of wastewater treatment facilities for petrochemical production facilities will not be reliable unless wastewaters have been fully characterized and the performance characteristics of alternative treatment processes have been evaluated by treatability studies and pilot plant operations. Treatability studies should establish the effects of operational parameters such as hydraulic detention time, sludge age and temperature on organic removal rates, oxygen requirements, sludge production, sludge characteristics, and process stability.

Treatability studies can identify wastestreams which should be treated separately to enhance process performance (Gloyna and Ford, 1979). Engineering-Science, Inc. (1971) outlined the components of a preliminary wastewater survey and treatability study in the petrochemical industry (Table 3.4). Biomass respiration measurements have been used to assess the treatability of petrochemical wastewaters. Respiration measurements as described by Somerville (1985), can be combined with chemical analysis to obtain a measurement of the potential oxidative capacity of an aerobic treatment system applied to a specified waste. The measurements obtained by these tests may be used to predict performance and aid in the control of processes. The wastewater treatability procedures used at one petrochemical plant have been described by Soderberg and Bockrath (1985).

Monitoring wastewater composition is an important part of water quality control and can be very complex and expensive. Several factors should be considered in a monitoring program. Sampling and/or monitoring locations should be selected to produce the most information per sampling effort. The accuracy required of the monitoring program should be determined. This will help determine how often samples must be taken and how large the samples must be. The parameters to be measured must be identified and suitable analytical procedures and instrumentation selected. The data collected should then be analyzed using the proper statistical techniques (Cheremisinoff, 1985).

Clements and Cheng (1982) described techniques which provide both qualitative and quantitative identification of major components of a process wastestream. The techniques and instrumentation used in this process were chosen to be within the technical and financial reach of even small company laboratories. Component identification was accomplished by using infrared spectroscopy, ultraviolet spectroscopy, gas chromatography, and thin-layer chromatography techniques. The sample preparation steps consist of an initial series of extractions which isolated compounds into organic acid, base and neutral compounds and a totally water soluble phase. Each fraction was then concentrated and subjected to the appro-

Table 3-2. Critical Precursor/Generic Process Combinations that Generate Priority Pollutants (Wise and Fahrenthold, 1981)

	Generic Processes				
Precursor	Oxidation products	Chlorination products	Nitration products	Diazotization products	Reduction products
Benzene	Phenol	Chloroaromatics Chlorophenols	Nitroaromatics Nitrophenols		
Toluene	o,m-Cresol		Nitroaromatics		
Xylene	2,4-Dimethyl-phenol		2,4-Dimethyl-phenol		
Naphthalene		2-Chloronaphthalene			
Phenol		Chlorophenols	Nitrophenols		
Cresols		4-Chloro-m-cresol	4,6-Dinitro-o-cresol		
Chloroanilines				Chlorophenols Chloroaromatics Aromatics	
Nitroanilines				Nitrophenols Nitroaromatics Aromatics	

Table 3-2. (continued)

Precursor	Generic Processes: Oxidation products	Chlorination products	Nitration products	Diazotization products	Reduction products
Nitrobenzene m-Chloronitro-benzene				N-Nitrosodi-phenylamine[a] Benzidines[b]	Aniline (diphenylamine[a]) 1,2-Diphenylhydrazines[b]
Ethylene		Chlorinated C2's Chlorinated C4 Chloroaromatics			
Propylene	Acrolein	Chlorinated C3's			
Methane		Chlorinated methanes			

a Derived directly from aniline, or indirectly via phenylhydrazine, diphenylamine is one of three secondary amines that are precursors for nitrosamines when exposed to nitrites (as in diazotization) or NO_x.

b Diphenylhydrazines rearrange to benzidines under acid conditions (as in diazotization).

Table 3-3. Petrochemical Processes (Hedley et al., 1975)

Cresol Formaldehyde Resins
Xylenes (separation)
Isophthalic Acid
Unsaturated Polyester (from Isophthalic Acid)
Phthalic Anhydride (from O-Xylene)
Phthalic Anhydride (from naphthalene)
Dinitrotoluene
Toluenediamine
Polyurethane
Maleic Andxdride (from benzene)
Unsaturated Polyester (from Maleic anhydride and phthalic anhydride)
Phenol (Toluene oxidation)
Phenol (Sulfonation of Benzene)
Phenol (from chlorobenzene)
Cumene
Phenol (cumene process)
Phenolic Resins
Bisphenol-A
Polycarbonate
Isopropanol (direct hydration)
Acetone (from isopropanal)
Methyl methaorylate (Cyanohydrin process)
Acrolein
Allyl Alcohol (from acrolein)
Glycerin (from Allyl Alcohol)
Allyl Chloride
Epichlorohydrin
Epoxy resin - a condensation of bisphenol-A and epichlorohydrin
Glycerin from Epichlorohydrin
Alkyd Resins
Polypropylene
Propylene Oxide (Chlorohydrin process)
Propylene Oxide (oxirane process)
Allyl Alcohol (from propylene oxide)
Acrylic Acid (propylene oxidation)
Ethyl Acrylate (direct esterification)
Acrylonitrile (aminoxidation of propylene)
Acetic Anhydride (from acetaldehyde)
Styrene Butadiene plastics
SAN Resins (mass polymerization)
ABS Resins (emulsion polymerization)
Butadiene (from n-Butylenes)
Ethylbenzene
Styrene
Polystyrene (suspension polymerization)
Polystyrene (bulk polymerization)
Ethylene Oxide (chlorohydrin process)
Ethylene Oxide (catalytic oxidation)

Table 3-3. (continued)

Ethylene Glycol (Hydration of ethxlene)
Polyethylene - High Density (Ziegler process)
Polyethylene - High Density (Phillips process)
Ethanol
Acetaldehyde (oxidation of ethanol)
Acetaldehyde (hydration of ethylene)
Acetic Acid (from acetaldehyde)
Acetic Acid (from methanol)
Low Density Polyethylene (autoclave reactor)
Low Density Polyethylene (tubular reactor)
Vinyl Acetate (from ethylene)
Vinyl Acetate (from acetylene)
Ethylene Dichloride (oxychlorination of ethylene)
Ethylene Dichloride (ethylene chlorination)
1,1,2-Trichloroethane (from ethylene dichloride)
Vinylidene Chloride (from 1,1,2-trichloroethane)
Vinyl Chloride (from ethylene dichloride)
Vinyl Chloride (from acetylene)
Polyvinylvinylidene Chloride
Polyvinyl Chloride
Polyvinvyl Acetate (emulsion polymerization)
Polyvinvyl Alcohol (hydrolysis)
Ethyl Acrylate (carbonylation of acetylene)
Methanol
Formaldehyde (from methanol)
Formaldehyde (from methane)
Amino Resins
Nitrobenzene
Aniline
Methylene Diphenyl diisocyanate (MDI)
Spandex
Cychlohexane
Cyclohexanone
Caprolactam (formation from hydroxylamine)
Nylon 6
KA Oil (from cyclohexane)
Adipic Acid
Adiponitrile (from adipic acid)
Adiponitrile (from butadiene)
Adiponitrile (from arylonitrile)
Nylon 66 (polymerization)
Hexamethylenediamine (from adiponitrile)
Cellulose Acetate
Terephthalic Acid (air oxidation)
Terephthalic Acid (nitric acid oxidation)
Polyethylene Teraphthalate (from terephthalic acid)
Polyethylene Teraphthalate (from dimethyl terephthalate)
Dimethyl Terephthalate
Nitroglycerine
Hexamethylene Tetramine

Table 3-3. (continued)

Diisodecyl Adipate
Tricresyl Phosphate
Cresyldiphenyl Phosphate
Tri phenyl Phosphate
Di-2-Ethylhexyl Phthalate
Ditridecyl Phthalate
Diisooctal Phthalate
Dimethyl Phthalate
Butyl Octyl Phthalate
Dibutyl Phthalate
Diethyl Phthalate
n-Octyl n-Decyl Phthalate
Diisodecyl Phthalate
Cyclooctadiene
Toluene Sulfonic acid
Toluene Sulfonate (hydrotrope)
Benzene Sulfonate (hydrotrope)
Xylene Sulfonic Acid
Xylene Sulfonate
Propylene Trimer and Tetramer
Dodecxlbenzene (hard alkylbenzene AB)
Cumene Sulfonic Acid
Cumene Sulfonate (Hydrotrope)
Nonylphenol (from propylene trimer)
Octylphenol (from butylene dimer)
Ethoxylated Nonylphenol
Ethoxylated Octylphenol (non ionic)
Butylene Dimer
n-Paraffin Chloride
Mixed Olefinic Product
OXO mixed Linear Alcohols
Ethoxylated Mixed Linear Alcohols (AEO)
Sulfated Ethoxylates (AEOS)
Alcohol Sulfates (sulfation with sulfur trioxide)
Alcohol Sulfates (sulfation with chlorosulfonic acid)
Linear Alcohols C_{12}-C_{16} (Ziegler process)
Linear Alkylbenzene (LAB)
Linear Alkylbenzene Sulfonate (LAS)
Mixed Linear Alcohols (oxidation)
Sym-trimethylene-trinitramine
Pentaerythritol
Pentaerythritol Tetranitrate
Polychloroprene
Butyl Rubber
Chloroprene (from Butadiene)
2-Methyl-1-Pentene
2-Methyl-2-Pentene
Isoprene (from 2-methyl-2-pentene)
Styrene-Butadiene Rubber (cold emulsion polymerization)
Styrene-Butadiene Rubber (hot emulsion polymerization)

Table 3-3. (continued)

Styrene-Butadiene Rubber (solution polymerization)
Modacrylic Fibers
Polyacrylonitrile (solution polymerization)
Polyacrylonitrile (suspension polymerization)
Polyacrylonitrile (emulsion polymerization)
Polysulfide Rubber (2 processes)
Nitrile Rubber
Ethyl Ether
Ethyl Acetate
Methyl Acetate
Isopropyl Acetate
n-Butyl Acetate
Acetic Anhydride (from acetic acid)
Acetic Anhydride (from acetic acid and ketene)
Isoprene (from amylenes)
Polyisoprene Rubber
Trinitro Toluene
Cellosolve
Ethyl Cellosolve Acetate
Crotonaldehyde
n-Butyl Alcohol (from crotonaldehyde)
t-Butyl Alcohol
Polybutadiene
Bis-2-Chloraethylformal
Methylene Chloride (from methane)
Trichloroethylene (from acetylene)
Perchloroethylene (from trichloroethylene)
1,1,1-Trichloroethylene
SEC-butyl Alcohol
Methyl Ethyl Ketone (from Sec-butyl Alcohol)
Methyl Isobutyl Ketone (from acetone)
n-Butyraldehyde (OXO reaction)
n-Butanol (n-butyraldehyde)
EPR Rubber (solution polymerization)
EPR Rubber (suspension polymerization)
EPT Rubbers
2-Ethyl Hexanol
Di-2-Ethylhexyl Adipate
Octyl Decyl Adipate

priate analytical technique. This procedure was found to be the most economical for wastestreams with contaminant concentrations in excess of 1000 mg/L, as is often found in petrochemical wastewaters.

The unit processes which can provide treatment of petrochemical manufacturing plant wastewaters are as varied as the unit processes used in the manufacturing plants themselves. Studies have shown, however, that there are seldom cost effective alternatives to biological treatment coupled with physical-chemical pretreatment and/or polishing where needed (Ford and Tishler, 1974; Gloyna and Ford, 1979; Nijst, 1978). Biological treatment coupled with postfiltration has been defined by the US EPA as the "best practicable technology" currently available for treating petrochemical processing wastewaters.

In order to produce a high quality effluent, it is probable that most petrochemical wastewater systems will include all or some of the processes listed in Table 3.5 (Ford and Tischler, 1974).

Nijst (1978) of the Petrochemicals/Ecology group of the European Council of

Table 3-4. Components of a Wastewater Survey and Treatability Study Program (Engineering-Science, Inc., 1971)

I. Wastewater Survey

   A. Identify all significant waste sources

   B. Obtain detailed information on waste flowrates

II. Wastewater Characterization

   A. Characterize organic content of wastestream

   B. Characterize inorganic content of wastestream

   C. Identify toxic wastestreams

   D. Identify wastestreams with reuse or product recovery potential

III. In-plant Considerations

   A. Implement educational programs for plant personnel designed to reduce wastewater generation

   B. Eliminate waste by process research and development

   C. Install waste segregation devices at source

IV. Treatability Study

   A. Select appropriate process alternatives based on wastewater survey and characterization data

   B. Define process operating parameter by bench or pilot scale process simulations

   C. Evaluate process alternatives based on treatment costs and treatment requirements

V. Incorporation of Results

Chemical Manufacturers Federation reports that a central biological treatment plant is the preferred method of treating the aqueous effluents of the petrochemical industry. Biological treatment processes were chosen because: 1) they are oriented toward BOD removal, which was generally required by the responsible authorities, 2) biological process costs to achieve BOD removal are low compared to other treatment processes, and 3) when effluent limitations are in terms of COD, biological processes will remove a significant amount of COD with less expense than a nonbiological process designed to remove the same amount of COD. This report further states that effluents from biological treatment systems may still contain dissolved organics and suspended solids which may be removed by further treatment such as aerated lagoons and polishing ponds, sand or multimedia filtration or other physical chemical processes such as reverse osmosis, ultra—filtration, extraction and chemical oxidation. Activated carbon adsorption preceded by filtration for solids removal was generally found to be the best economically available technology for reducing the residual COD of biologically treated effluents (Nijst, 1978).

Table 3-5. Petrochemical Wastewater Treatment Systems (Ford and Tischler, 1974)

Pretreatment:

- API separators
- Tilted plate separators
- Filtration for oil removal
- pH control
- stripping processes
- primary sedimentation

Intermediate treatment:

- Dissolved air flotation
- coagulation-precipitation
- equalization

Secondary/Tertiary treatment:

- Biological oxidation
- Chemical oxidation
- Filtration
- Adsorption

Kulperger (1972) used a high purity oxygen activated sludge system for treating petrochemical wastes. The system consisted of a four stage biological reactor and a center feed clarifier. Pilot plant studies showed that the system could provide a BOD removal of 90% in a wastestream which contained 2,700 to 4,000 mg/L BOD. The plant was operated at biomass loadings of 0.49 to 0.82 pounds of BOD_5 per pound of MLVSS and required 1.6 pounds of oxygen to remove one pound of BOD at a loading rate of 0.6 pounds of BOD per pound of MLVSS per day.

Biological unit operations at petrochemical producing plants may be subjected to shock loadings of toxic organics, such as those described in the "Wastewater Characteristics" section of this chapter. For this reason, biological wastewater treatment systems should be designed with long sludge retention times (Rozich and Gaudy, 1985). The use of equalization basins to minimize the effects of these toxic shock loads is also recommended (Dold et al. 1985).

Sequencing batch reactors (SBR) have also been used to treat petrochemical wastewaters. These systems have several advantages (Mandt, 1985). Inflow into the system and hydraulic residence time are easily controlled. Finally, very long sludge detention times may be maintained in these units. SBR plants have successfully treated wastestreams contaminated with phenols and other toxic substances (Herzburn et al. 1985).

Mixing petrochemical wastewaters with domestic wastewaters has been found to be beneficial in some cases. Effective treatment, including nitrification, was accomplished by an activated sludge system when wastewaters from oil refineries, petrochemical plants and domestic wastewaters were mixed prior to treatment (Nesaratnam and Ghobrial 1985). When the same wastewater was treated in a fluidized bed process, accumulation of oil inhibited system performance.

The utility of anaerobic lagoon pretreatment of petrochemical waste was investigated by Hovious et al. (1973). A design procedure for the selection of lagoon volume based on organic loading and temperature was presented. Using this design procedure it was estimated that a lagoon with a hydraulic detention time of about 10 days and a temperature of 20°C would achieve 40 percent COD removal and 50 percent BOD removal when the influent contained 3,000 mg/L of COD. Examination of chromatographically identifiable organic compounds in the waste used during this research indicated that all compounds, except metabolic intermediates, were removed to a significant degree in the anaerobic lagoon (Table 3.6).

Britz et al. (1983) reported the successful use of downflow fixed bed anaerobic reactors for the treatment of a petrochemical effluent. COD reductions of 93-95% were found at an optimum retention time of 2.3 days and a loading rate of 4.7 kg $COD/m^3/d$. Approximately 0.88 $m^3/m^3/d$ (at standard temperature and pressure) of biogas was produced with a methane content of 90-96%.

Fisher et al. (1971) also investigated the use of anaerobic processes for petrochemical waste treatment. Packed bed, mixed digester and anaerobic lagoon processes were evaluated. The anaerobic lagoon was found to be the process of choice. Investment and operating costs were the lowest of the studied systems, and a microbiological sulfur reduction-oxidation cycle occurred in the lagoon in which sulfates were used and organics removed. The anaerobic system produced smaller amounts of biomass and required less energy for operation. Some compounds were degraded in the anaerobic system which could not be aerobically degraded.

An acidic effluent from a petrochemical plant was treated using an upflow anaerobic sludge blanket reactor. The system operated best at a hydraulic retention time of 1.78 days and a loading rate of 7.255 kg $COD/m^3/day$. At these

Table 3-6. Removal of Specific Organics in Anaerobic Lagoons* (Hovious et al., 1973)

Compound	Dilute Wastes: Loading Rate 13 lb COD/day/ 1,000 cu ft		Concentrated Wastes: Loading Rate	22 lb.COD/day/ 1,000 cu ft	48 lb.COD/day/ 1,000 cu ft
	Influent (mg/L)	Effluent (mg/L)	Influent (mg/L)	Effluent (mg/L)	Effluent (mg/L)
Methanol	80	35	380	135	145
Ethanol	80	15	270	120	130
n-Propanol	—	—	170	35	40
Isopropanol	60	30	175	45	55
n-Butanol	—	—	170	75	80
Isobutanol	—	—	250	80	85
n-Pentanol	—	—	315	70	100
Isopentanol					
Hexanol	—	—	140	20	30
Acetaldehyde	30	10	80	35	40
n-Butyraldehyde	—	—	190	50	35
Isobutyraldehyde	—	—	210	50	50
Acetone	90	60	150	80	70
Methylethyl ketone	10	5	---	---	---
Benzene	10	5	---	---	---
Ethylene glycol	135	30	755	155	190
Acetic acid	215	220	2,120	2,280	2,620
Propionic acid	—	—	0	505	470
Butyric acid	—	—	0	330	300

*Data are averaged from 5 to 12 occurrences in grab or composite samples.
Note: Lb/day/1,000 cu ft x 16 = g/day/cu m.

conditions, 83% of the COD was removed with 2.64 m^3/m^3/day of gas produced which contained less than 90% methane (Nel et al. 1984).

Temperature effects on the biological treatment of petrochemical wastewaters were investigated by del Pino and Zirk (1982). Empirical models were developed to fit the relationship between effluent BOD and COD and hydraulic retention time, mixed liquor volatile suspended solids, temperature and influent substrate concentration. The effects of temperature on the biological treatment of petrochemical wastewaters were observed to be more drastic than temperature effects on municipal wastewater treatment systems. Tests to measure the effects of temperature on sludge characteristics were inconclusive.

The biological treatment of a complex petrochemical wastestream using a sequence of anaerobic digestion and activated sludge was studied by Humphrey et al. (1979). Bench scale and pilot plant studies using various composite samples and process wastewater blends indicated the need for stream segregation and waste reduction. The system was effective in removing the biodegradable portion of the pretreated wastewater stream. The average influent composition of approximately 6000 mg BOD/L, 8000 mg COD/L and 1000 mg nitrates/L was reduced to an effluent with approximately 50 mg BOD/L, 1200 mg COD/L, 200 mg suspended solids/L and essentially no nitrates. These data show that a significant amount of COD could not be removed by conventional biological treatment processes.

Studies conducted by Medley and Stover (1983) and Stover et al. (1983) have shown that pretreatment with ozone can increase the biodegradation of some organic compounds found in petrochemical wastewaters. Ozone addition was found to be beneficial; however, it is not a "cure-all" and studies should be conducted on each compound to determine if it is effective and economical. The addition of powdered activated carbon (PAC) to biological oxidation processes may also significantly enhance the efficiency of biological treatment processes (Rolling et al. 1983).

Recent advances in the field of bioengineering have lead to the development of microbial cultures which have the ability to break down molecules resistant to biological degradation. Thibault and Zitrides (1979) have reported that a specially adapted strain of bacterial inoculum applied to the biological treatment process at a petrochemical processing plant significantly improved effluent quality. The addition of these selectively adapted microbes reduced effluent total oxygen demand, and biochemical oxygen demand, improved system stability, eliminated an existing foam problem, and resulted in the elimination of at least one compound (tertiary butyl alcohol) from the effluent which was not previously degraded. These results, combined with recent advances and interest in genetic engineering, suggest that biological treatment processes may be improved by these techniques; however, further research is required.

Physical-chemical processes play an important role in petrochemical wastewater treatment. Many physical-chemical treatment processes are used to pretreat petrochemical wastewater in preparation for biological treatment. API separators are used to remove materials less dense than water, such as free oil, and suspended matter that is more dense than water. Tilted plate separators are also used to remove materials less dense than water. Several types of filtration devices are also used to remove free oil and solids from wastestreams prior to biological treatment.

Neutralization is commonly required in the treatment of petrochemical wastewaters. Acid streams may be neutralized by fluidized mixing with lime slurries, dolomitic lime slurries, caustic or soda ash. Alkaline streams may be neutralized with sulfuric or hydrochloric acid or with boiler flue gas (carbon dioxide). Neutralization can often be accomplished by mixing internal wastewater streams (Ford and Tischler, 1974). Volatile organic compounds, hydrogen sulfide and ammonia are often removed from wastewater streams by stripping processes.

Dissolved Air Flotation (DAF) is commonly used in petrochemical waste treatment plants to enhance oil and suspended solids removal. DAF units, while not as economical as API separators and tilted plate separators, produce a better quality effluent which is often required to meet effluent oil limitations. If a significant portion of the oil is emulsified, chemical addition with flocculation chambers may be a part of the flotation unit. Coagulation-flocculation processes are effective in removing suspended solids, some nutrients and heavy metals from petrochemical wastestreams (Ford and Tischler, 1974).

Activated carbon adsorption systems are often used to remove residual organic compounds from both aqueous and gaseous wastestreams from petrochemical plants. Compounds in the alcohol, aldehyde, amine, pyridine, morpholine, aromatic, ester, ether, glycol, glycol ether, ketone, organic acid, oxide and halogenated organic groups have been found to be amenable to carbon adsorption (Giusti et al. 1974). Carbon adsorption studies conducted using municipal/industrial wastewater have also shown that granular activated carbon is an effective treatment method for the removal of the US EPA's "priority pollutants" (McManus et al. 1985). The activated carbon process has been well studied and is outlined in many wastewater treatment texts and other handbooks (US EPA, 1973; Tchobanoglous and Schroeder, 1985). A review of the literature on activated carbon adsorption as a treatment concept used in the petrochemical industry was presented by Matthews (1978).

Petrochemical wastewater was treated in a treatment system consisting of oil removal, biological oxidation, chemical treatment, filtration and activated carbon adsorption. The COD was reduced from 3,200 mg/L to 30 mg/L. The activated carbon columnswere found to remove dissolved organics not amenable to biological treatment along with color (Christensen and Conn, 1976).

Four physical-chemical unit operations were studied by Coco et al. (1979) to determine their feasibility for removing biorefractory organics found in petrochemical wastestreams. Steam stripping was evaluated using petrochemical process effluents containing chlorinated hydrocarbons and aromatic hydrocarbons. This unit operation removed up to 75 percent of the total organic carbon (TOC) in the process effluent. The cost of this treatment process was significantly reduced by the recovery of lost product.

Solvent extraction was evaluated using process effluents containing chlorinated hydrocarbons and aromatic hydrocarbons. Straight chain paraffin hydrocarbons in the C_{10} to C_{12} range were found to give maximum TOC removal with minimum TOC residual. Organic removals in the 90 percent range were frequently obtained during pilot plant operation (Coco et al. 1979). Product present in the wastestream was also recovered in this process, and thus contributes to reduced treatment costs.

Ozonation was an effective method of pretreating wastewaters from the manu acture of toluene di-isocyanate, ethylene glycol, styrene monomer, and ethylene dichloride. Batch oxidation studies showed that ozonation improved biotreatability of these wastewaters. Complete oxidation with ozone was found to be uneconomical (Coco et al. 1979). Carbon adsorbtion removed C_1 and C_2 chlorinated hydrocarbons from the wastestreams studied. Adsorption characteristics of different commercially available activated carbons were evaluated. In addition, an activated carbon was developed from a by-product soot produced in the acetylene process. This carbon was found to have about 80 percent of the absorptive capacity of commercial products (Coco et al. 1979).

Other processes which have been used for the treatment of petrochemical wastewaters include: polymeric adsorption (Fox, 1975; Vinson, 1972), wet air oxidation (DeAngelo and Wilhelmi, 1983; Rappe, 1985; Canney et al., 1985), pyrolysis, (ChemicalEngineering, 1982) and free radical oxidation (Feuerstein, 1982; Legan, 1982; Khan et al. 1985).

Process Modification, Conservation and Treatment

The petrochemical industry lends itself to controlling pollution through process improvement rather than pollution abatement. Four alternative possible solutions may be developed for a pollution problem in the petrochemical industry depending on the waste produced. First, some wastes may be recovered as salable coproducts. Second, wastestreams can be recycled after some process modification for conversion to prime product or for reuse in the process as a reagent or intermediate. Third, the waste may be usable as a fuel. Fourth, and least desirable, wastes may be treated in waste treatment processes where they are converted to less harmful states and/or dispersed in quantities which may be assimilated by the environment.

Summaries of some process improvements which have aided in the reduction of pollutants in the petrochemical industry may be found in Burroughs (1963), Mencher (1967), the *Oil and Gas Journal* (1967), and Rickles (1965).

Process technology in the petrochemical production industry is constantly changing. Some unit processes will produce desired products with a reduction in the quantity of pollutants generated when compared to other technologies. Tavlarides (1983, 1985) developed a matrix of significant pollution problems and process modifications which will reduce or, in some cases, eliminate the production of these pollutants for the explosives industry and other industries. Matrices for nitric acid, TNT, and nitrocellulose production are shown in Tables 3.7 through 3.9, respectively. The matrices describe individual processes, the pollutants and their sources in the process, the nature of the pollutant and the process modification for mitigation or reduction of the pollutant. Analysis of production unit processes may, therefore, lead to the production of smaller amounts of water-borne pollutants.

Several schemes were suggested by Quartulli (1975) to reduce water consumption and increase the use of waste streams as process raw materials in steam-hydrocarbon reforming plants. The processes proposed recycling essentially all of the process condensate to the process system (with minimum offsite treatment) and bypassing feedwater, boiler and steam turbine systems.

Petrochemical wastewaters containing high concentrations of salt and refractory organic contaminants were treated by activated carbon adsorption for the removal of organic constituents (Zeitoun, 1979). The remaining salt solution was treated with a hybrid electrodialysis-reverse osmosis process to produce freshwater and a concentrated brine solution. Organics recovered by the system were recycled and the processed water was suitable for reuse.

Conserving and reusing water have become key concerns in chemical processing industry plants as the availability and quality of water supplies diminish and wastewater discharge regulations become more stringent. Holiday (1982) has discussed the use of eight technologies which may be applied to reduce water usage by either cutting usage at some point in the plant or by recycling and reusing a waste stream. The technologies described include vaporcompression evaporation, waste-heat evaporation, reverse osmosis and ultrafiltration, electrodialysis, steam stripping, combination wet/dry cooling towers, air-fin cooling and cooling water sidestream softening. A list of the characteristics of these processes is contained in Table 3.10 (Holiday, 1982). The use of reverse osmosis as a water purification process in the petrochemical industry was reported by Kosarek (1979).

Willenbrink (1973) reported the use of several techniques to reduce or concentrate wastestreams containing phenol, ammonia and hydrogen sulfide. In one process, the use of a dilute caustic in product purification washes drastically

Table 3-7. Pollutants and Control Options in Nitric Acid Production Process (Tavlarides, 1983)

Process	Pollutants	Source in Process	Nature of Pollutants	Pollutant Control Strategy
Ammonia Oxydation	NO_x, N_2	Tail gas from absorption tower	Inorganic gases	Adjustment of operating temperature for maximum HNO_3 output. Mixing of gases at the inlet of catalytic bed. Improvement of mass transfer rate between the catalyst and the bulk gas.
Nitric acid concentration	NO_x	Absorption tail gas	Inorganic gases	Provide sufficient air supply. Prevention of leakage from ammonia oxidation process.
Spent acid recovery	Nitrogen salts nitrobodies	From nitration processes	Organic and inorganic	Removal of suspended solid

Table 3-8. Pollutants and Control Options in TNT Production Process (Tavlarides, 1983)

Process	Pollutants	Source in Process	Nature of Pollutants	Pollutant Control Strategy
Nitration	CO, CO_2, NO, N_2O, NO_x, Trinitromethane (TNM) unsymmetrical "Meta" isomers of TNT	Nitrator-Separator	Inorganic and organic, gases	Low temperature dinitration stage to reduce "meta" isomers and oxides of nitrogen and carbon.
		Fume recovery system	Inorganic and organic gases	Change the design as to operate at maximum TNT
	"Yellow water" acidic effluent traces of TNT	First water wash	Inorganic and organic liquid	Removal of TNT from all waste waters by application of foam separation. "Red water" generation will be considerably lower when "meta" isomers formation is decreased.
Purification	"Red water" $NaNO_x$, Na_2SO_x nitrotoluenes and nitrotoluenes sulfuric acid	Sellite wash	Inorganic and Organic liquids	
	"Pink Water" DNT, TNT, all pollutants present in red water	Nitration fume scrubber discharge "Red water" concentration distillates. Finishing operation hard scrubber Wash down effluent	Inorganic and organic liquid	
Finishing	Spillage, drainage.			

Table 3-9. Pollutants and Control Options in Nitrocellulose Production (Tavlarides, 1983)

Process	Pollutants	Source in Process	Pollutant	Control Strategy
Nitration	SO_x, NO_x, NO_3 suspended solids	Reactors nitration vessel	Inorganic gases and solids	Improve reaction by temperature adjustment to reduce generation of oxides.
	Waste acid HNO_3, H_2SO_4	Centrifuge	Inorganic acids	Conversion to sulfuric acid by spraying acid waste by water in presence of roasting gases containing SO_2.
Purification	Nitrocellulose fines	Boiling Tub house Jordan beater house poacher house	Organic, suspended solids	Removal by filtration from waste streams.
	Process water, acidic.		Inorganic acids	Reduction of process water.

Table 3-10. Conserving and Reusing Water; Water Conservation and Reuse Technologies that are Seeing Wider Application (Holiday, 1982)

Technique	Application	Limitations	Relative costs: Capital	Relative costs: Operating	Comments
Vapor-compression evaporation	Concentration of wastewater or cooling-tower blowdown Concurrent production of high-purity water	Not for organics that form azeotropes or steam-distill Fouling must be controllable	High	High	Rapid growth High-quality distillate Handles broad range of contaminants in water
Waste heat evaporation	Concentration of wastewater Condensate recovery	Not for organics that form azeotropes or steam-distill	Medium	Medium	Not widely used now Future potential good
Reverse osmosis, ultrafiltration	Removal of ionized salts, plus many organics Recovery of heavy metals, colloidal material Production of ultrapure water	Fouling-sensitive Stream must not degrade membranes Reject stream may be high-volume	Medium	Medium	Future potential strong Intense application development underway
Electrodialysis	Potable water from saline or brackish source	Limited to ionizable salts	Medium-high	Medium	Modest future potential

Table 3-10. (continued)

Technique	Application	Limitations	Relative costs: Capital	Relative costs: Operating	Comments
Steam stripping	Recovery of process condensates and other contaminated waters Removal of H_2S, NH_3, plus some light organics	Stripped condensates may need further processing	Medium	Medium-high	Well-established as part of some processes
Combination wet/dry cooling towers	Puts part of tower load on air fins Can cut fogging	Costly compared with wet cooling tower	Medium	Medium	Growth expected in arid areas
Air-fin cooling	Numerous process applications	For higher-level heat transfer Can be prone to freeze-up, waxing	Medium	Medium	Well-established Good for higher-temperature heat rejection
Sidestream softening	Reduce cooling-tower blowdown	Dissolved solids must be removable Control can be difficult	Low-medium	Low-medium	Not widely used Future potential good

reduced the amount of phenol in the wash water wastestream. This system not only reduced the amount of phenol to be treated, but also reduced the consumption of caustic with no noticeable product deterioration. Steam stripping regenerated cataylst was used to minimize the introduction of oxygen into a fluid catalytic cracking operation to reduce the production of phenolic compounds. Wash waters and steam condensates which have been in contact with hydrocarbons were collected from various unit operations for use as wash water to prevent corrosion and salt build up. A fraction of the phenol present in this washwater is absorbed by the hydrocarbons being washed. Reductions of approximately 50% of the phenol present have been observed.

Several water reuse and recycling systems currently used in the petrochemical industry were described by Dennis (1979). The processes include separation of potable and process water, recycled non-contact cooling water systems, process water recycle, and spray irrigation of process wastewater. A computer monitored and controlled system designed to manage water and energy at petrochemical plants was developed by Kempen (1982). This system helped reduce water use, wastewater treatment costs and energy use.

Case Histories

Wastewater from a petrochemical plant which produces raw materials for the polyester fiber and film and polystyrene industries are treated by an activated sludge plant consisting of equalization basins, biological oxidation basins, clarification, dissolved air flotation, polishing ponds and filtration (Figure 3.1). The waste treatment facilities are described in Table 3.11 (Chodil, 1983). This facility produced an effluent which contained 8 mg/L BOD and 11 mg/L suspended solids while removing 98% of the TOC present. A unique management strategy was employed at the facility. Each production unit is held accountable for the wastes they generate. Waste treatment costs are charged back to the individual production units based on the amount of organics discharged to the treatment plant. This system creates incentives for personnel at each unit to reduce waste loads and product losses.

Ford et al. (1973) described the development of a water pollution control system for the Zulia El Tablazo Petrochemical Complex in Venezuela. The first task in designing the treatment system was to define each of the production processes and predict the qualitative and quantitative characteristics of the wastewaters. Treatability studies were then performed to evaluate treatment alternatives. Design criteria were developed from the results of treatability tests and conceptual flowsheets were developed.

An activated sludge system was deemed to be the most practical, reliable and economical method of treating the El Tablazo wastewaters to the desired level. This decision was made based on treatability study data and the experience of a consultant in previous investigation, design, and operation of petrochemical plant wastewater treatment systems (Ford et al. 1973).

A simplified schematic of the treatment system is shown in Figure 3.2. The system was devised so that all dry weather organic sewer flow is treated. An impoundment basin was constructed for temporary storage and controlled release of specific wastewaters to the treatment process. The system also includes an equalization basin to minimize hydraulic and waste load variations. The equalized flow enters parallel activated sludge basins. Additional parallel units may be

Table 3-11. Wastewater Treatment Plant Data Sheet (Chodil, 1983)

Utilizes activated sludge process for treating high strength organic acid wastewaters before final discharge to the Des Plaines River.

Facilities

	Capacity	Detention time
Feed Equalization & Surge	15 million gallons	30 days
6 Aeration Basins	3 million gallons	6 days
2 Secondary Clarifiers	188,000 gallons	
Dissolved Air Flotation Unit	59,000 gallons	
Polishing Lagoon	40 million gallons	60 days
Dual Media Filter	1,000 gpm	
Sludge Storage	3.5 million gallons	4 months
Sludge Landfarm	200 acres	

Manpower

Operators	1 per shift
Sludge hauling operator	1 per day shift
Foreman	1 per shift
Supervisor	1
Engineer	1
Lab technician	1
Maintenance	3 per day shift

Raw wastewater

	Average	Capacity
Flow, gal/day	500,000	700,000
Concentration, mg/L TOC	3,000-4,000	
TOC Load, lbs/day	14,000	24,000
COD Load, lbs/day	35,000	60,000
BOD_5 Load, lbs/day	25,000	43,000
Population Equivalent	150,000	250,000

Effluent quality

BOD_5, mg/L	8
Suspended Solids, mg/L	11
TOC Reduction, %	98

Costs

Total capital expenditure to date	US $9 million (1982 US dollars)
Annual operating, maintenance, and fixed costs	US $2.5 million
Treatment cost, per pound of TOC	US $0.50

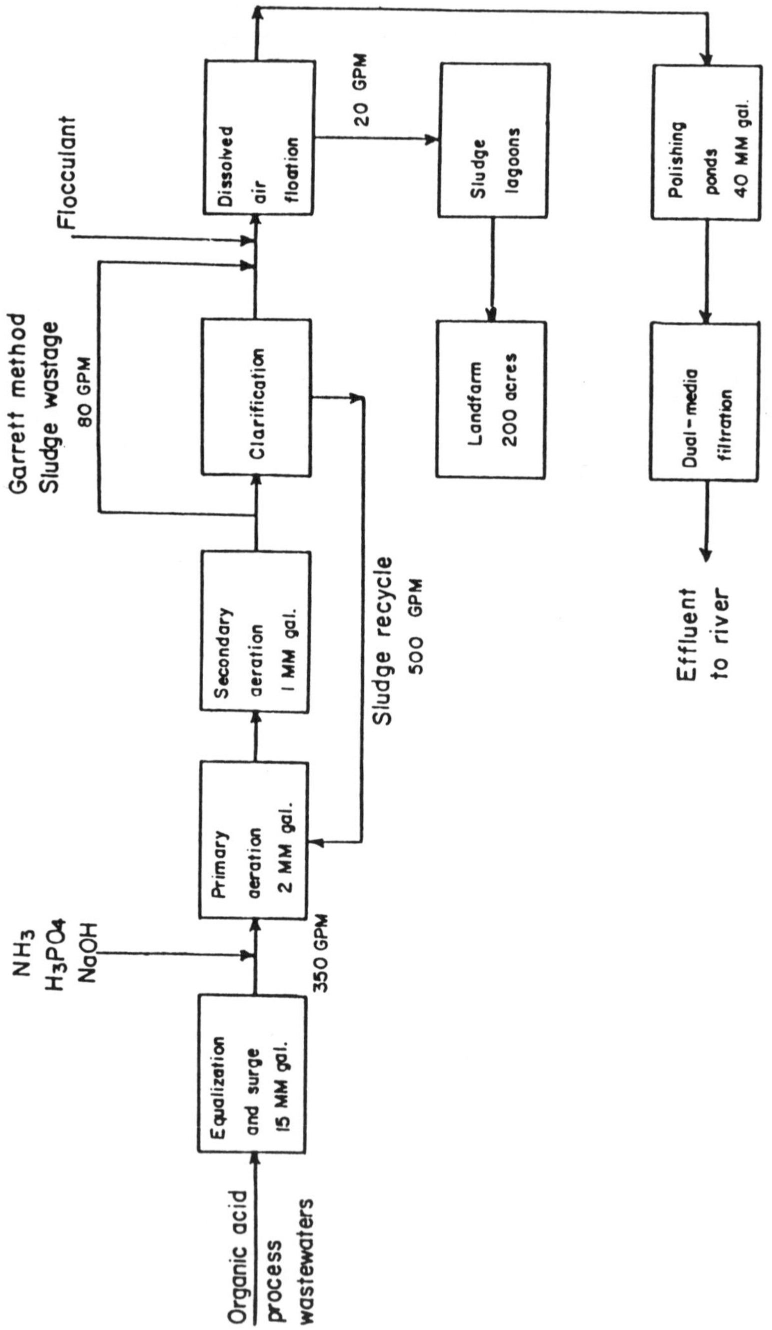

Figure 3-1. Waste Treatment Flow Diagram (Chodil, 1983)

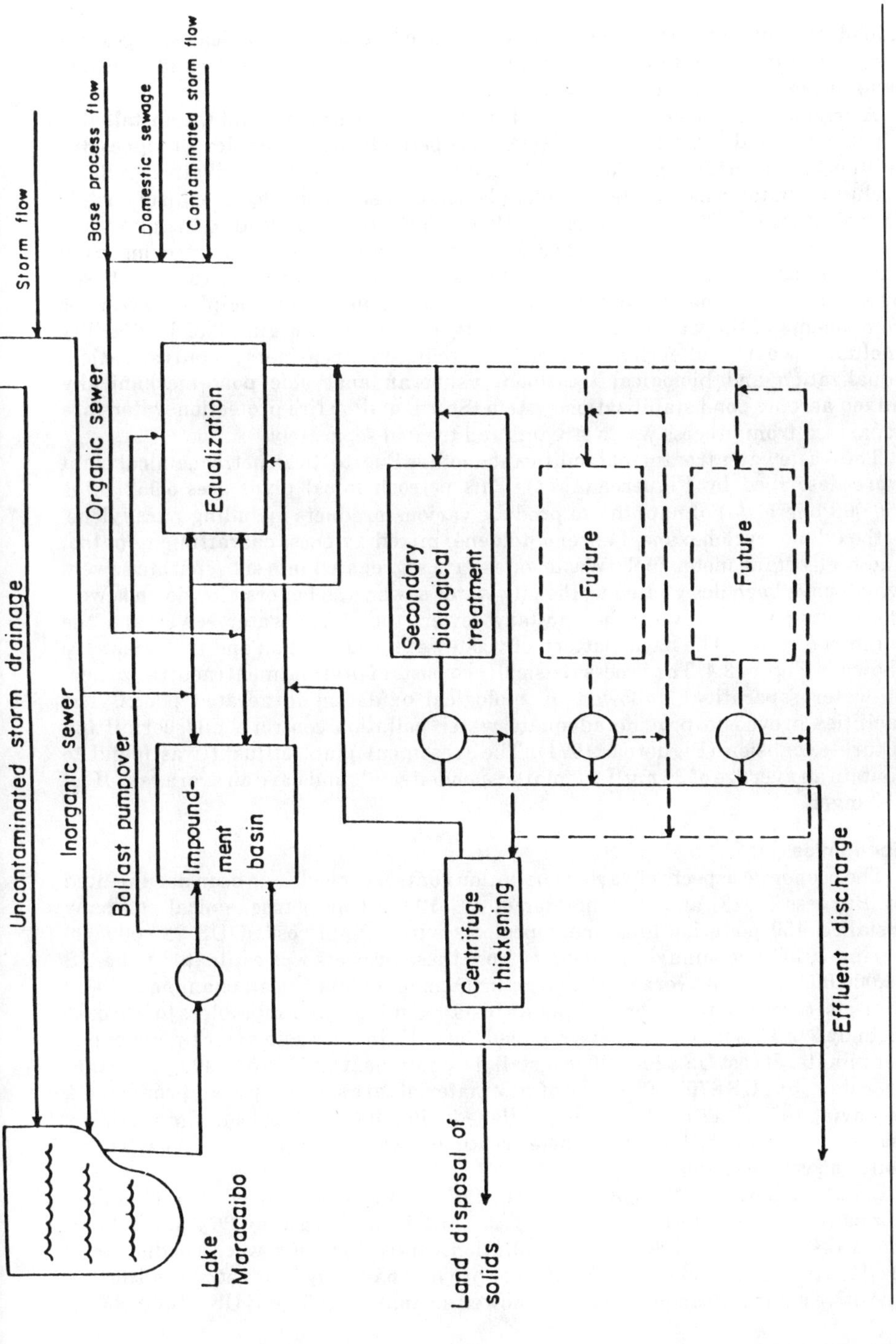

Figure 3-2. El Tablazo Treatment Complex (Ford et al., 1973)

added as more treatment capacity is required. Excess biological sludges are aerobically digested, thickened, dewatered by basket centrifugation and hauled to land disposal sites (Ford et al. 1973).

A primary and secondary biological wastewater treatment plant was installed at a petrochemical plant in Puerto Rico. The petrochemical complex produces 352 million kilograms (775 million pounds) of ethylene per year and derivative products including butadrene, ethylene oxide, phenol, cumene, polyethylene, bisphenol-A, and plasticizers (Rucker and Oeben, 1970). The total waste load to the plant was based on flow measurements and laboratory analyses of waste streams from existing plants plus estimates of aqueous waste loads from new process units based on an understanding of process chemistry and engineering principles. The major components of the waste treatment facility are shown in Figure 3.3. The facility includes waste collection equipment, primary treatment, neutralization, equalization and biological treatment using an anaerobic pond-mechanically mixed aerobic pond stabilization system. Storm and/or fire protection waters are separated from process waste streams and treated separately.

The wastewater treatment facilities at another Puerto Rican petrochemical plant were described by Figueroa (1971). This petrochemical plant uses 5,962 m^3/d (50,000 barrels/d) of naphtha to produce various products including paraxylene, orthoxylene, cyclohexane, benzene, toluene, mixed xylenes, parraffinic naphtha, and high octane motor fuel. Wastewaters are segregated into six separate sewers which have been designated as the oily water sewer, the boilers blowdown sewer, the cooling water sewer, the sanitary sewer, the storm water sewer and the combined sewer. The basic pattern of wastewater segregation and treatment are shown in Figure 3.4. The process basically consists of pretreatment (neutralization, oil-water separation), followed by biological oxidation in aerated ponds. The facilities proved to provide adequate water pollution control while permitting future expansion (Figueroa, 1971). The treatment plant effluent was found to contain an average of 25 mg/L of total suspended solids and have an average COD of 150 mg/L.

Economics

The economic aspects of various pollution control projects has been documented by Burgess (1973) and Eckenfelder et al. (1985). One petrochemical company installed 450 pollution abatement projects with a total cost of US $20,000,000 during 1971. The annual net savings from these projects was estimated to be US $6,000,000, with an average return on investment of 30%. As an example of these projects, three organic chemical plants installed mixers and subcoolers to increase yield by 3%. This process change reduced the COD in the wastewater by 30 pounds per minute. It cost US $200,000 to install the equipment and US $42,000 per year to operate it, but US $709,000 worth of raw materials are saved per year, producing a net savings of US $667,000 per year while reducing the pollution load. Emphasizing waste prevention rather than waste treatment was shown to be a cost effective pollution control strategy.

Pollution control efforts of one division of the above mentioned petrochemical company reduced BOD discharges by 25%, COD discharges by 26% and soluble solids discharges by 72%. Ninety pollution control projects were conducted to produce this reduction. Fifty-five of these projects had a negative return and 35 had a positive return. Economic analysis showed an annual savings of US $2,960,000 on

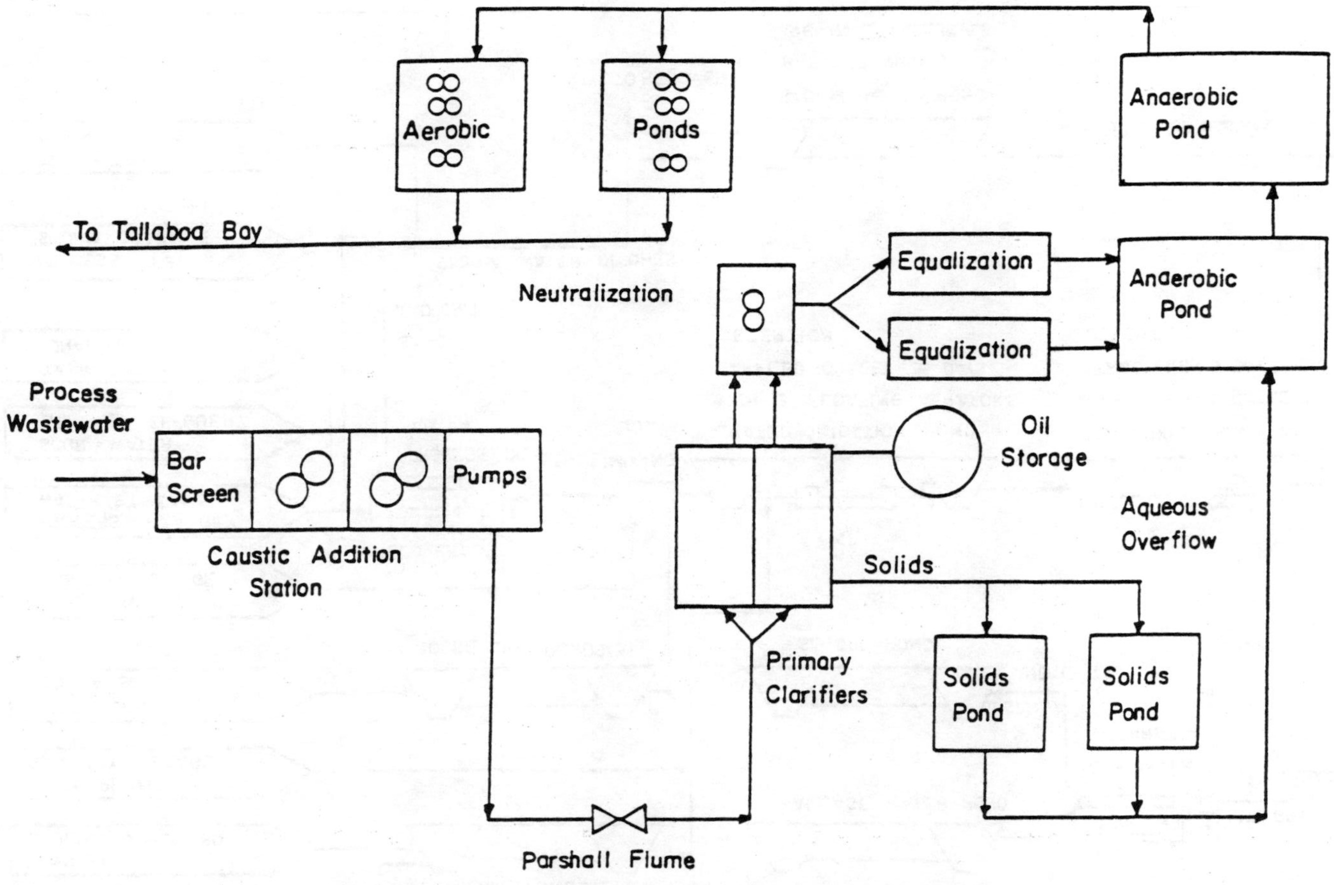

Figure 3-3. Flow Diagram of Wastewater Treatment Facility at Ponce, Puerto Rico (Rucker and Oeben, 1970)

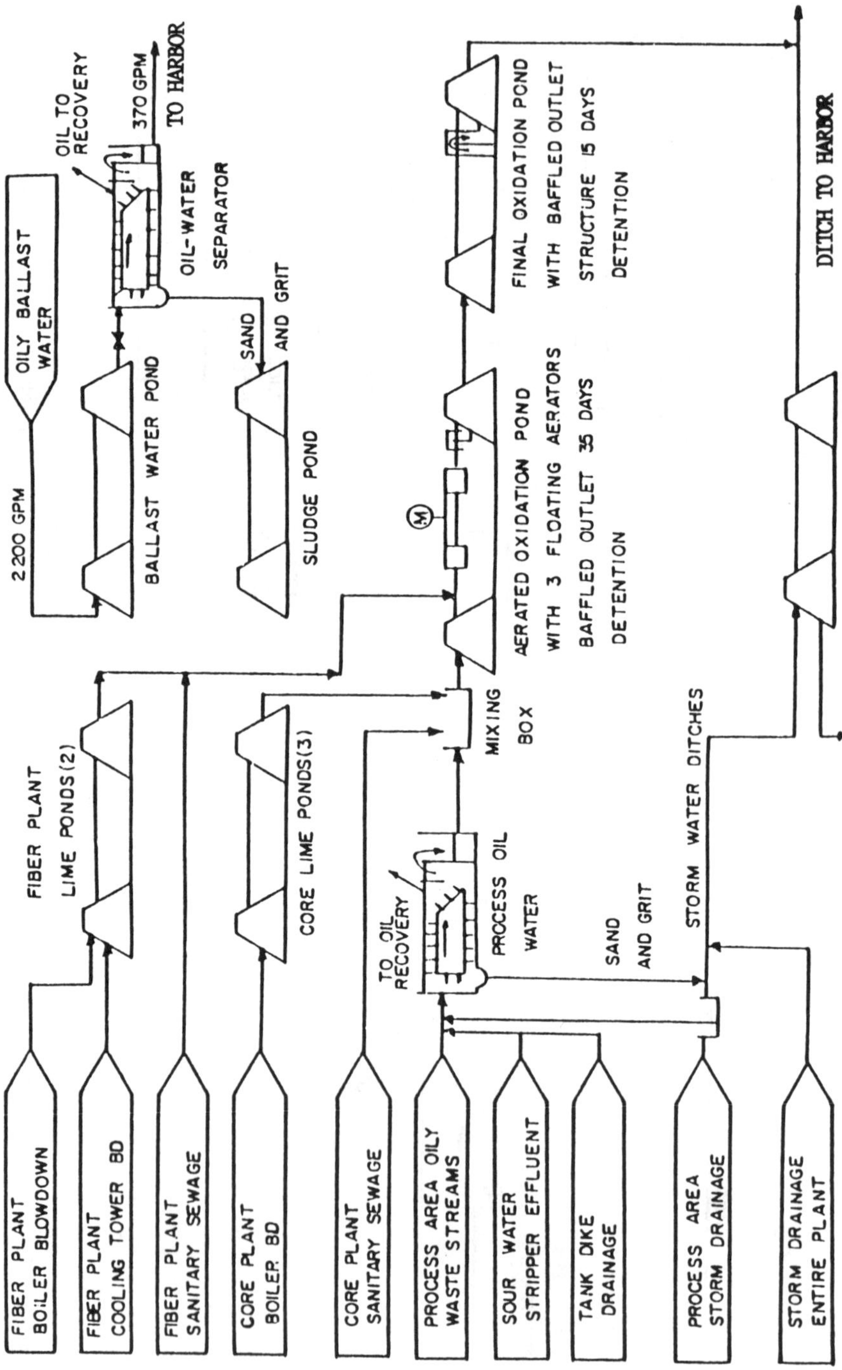

Figure 3-4. Simplified Flow Sheet, Wastewater Treatment, Guayama Petrochemical Complex (Figueroa, 1971)

a US \$6,100,000 investment (1971 dollars). An US \$8,300,000 capital investment produced a net annual savings of more than US \$1,000,000 in another division of this company, while an investment of US \$2,100,000 produced a US \$1,300,000 yearly savings in still another division (Burgess, 1971).

A flow measurement system coupled with computer control reduced the cost of water, steam generation, and steam distribution and produced less waste at a Belgian petrochemical plant (Kempen, 1982). The computer control system provided water and energy management practices which reduced energy requirements and water and wastewater treatment costs at the plant. The total project cost was approximately US \$435,000 (1982) and was projected to produce a return on investment of better than 64 percent.

A polymeric adsorption system efficiently removed and recovered phenol and was also found to produce a net savings while eliminating a waste treatment problem (Fox, 1975). In this process phenolic compounds may be removed from a waste-stream by adsorption onto a polymeric resin. During resin bed regeneration, the phenolic compounds may be recovered. In one application, it was estimated this system would produce a net savings of US \$235,000 per year, while producing an effluent with less than 1 mg/L of phenol.

Chlorinated and aromatic hydrocarbons were removed from petrochemical process effluent by steam stripping at a cost of US \$0.00041 per liter of treated wastewater including product recovery credit. Solvent extraction was also used and the cost of treatment, including product recovery credit, was estimated to be US \$0.0007 per liter of treated wastewater. Ozonation was an effective treatment method, but costs were much higher than those observed for steam stripping and solvent extraction. The operating costs for a carbon adsorption system removing organics from an ethylene dichloride plant effluent were estimated to be US \$3.50 to 4.50 per kilogram of organics removed or US \$0.045 to 0.07 per kilogram of product (Coco et al. 1979).

References

1. Bridie, A. L., J. M Wolff and M. Winter. 1979. BOD and COD of some petrochemicals. *Water Research*, 13 (17):627.
2. Britz, T. J., L. C. Meyer and P. J. Botes. 1983. Anaerobic digestion of a petrochemical effluent. *Biotechnology Letters*, 5(2):113.
3. Burgess, K. L. 1973. Clean up environmental problems in the petrochemical and resin industry can be profitable. In: *The petroleum/petrochemical industry and the ecological challenge.* American Institute of Chemical Engineers. New York, NY.
4. Burroughs, L. C. 1963. Recent developments in control of air and water pollution in U.S. refineries. In: *Proceeding of Sixth World Petroleum Congress,* Frankfurt.
5. Canney, P. J., et al. 1985. PACT wastewater treatment for toxic waste cleanup. *Proceedings of the 39th Industrial Waste Conference, 1984, Purdue Univ.* Ann Arbor Science Publishers, Inc., Ann Arbor, MI.
6. *Chemical Engineering.* 1982. 89(25):41. Feuerstein, W. 1982. The application of Fentons Reagent for chemical oxidation in wastewater treatment. Kern forschungszent. Karlsruhe, [Ber.] KFK (Ger.), KFK 3274; *Chemical Abstracts*, 96, 204-845.
7. Cheremisinoff, P. N. 1985. Special Report: Monitoring of Wastewater and Water Quality. Pollution Engineering 17 (9):40.
8. Chodil, G. P. 1983. Wastewater treatment an award winning design. *Hydrocarbon Processing*, 62(10):61.
9. Christensen, D. R. and B. R. Conn. 1976. Advanced wastewater treatment process is effective. *Hydrocarbon Processing*, 55(10):107.
10. Clements, L. D. and S. W. Cheng. 1982. Characterization of hydrocarbon pollutant bordens in petrochemical and refinery process streams. Environment International, 7(4):259.
11. Coco, J. H., E. Klein, D. Howland, J. H. Mayes, W. A. Myers, E. Pratz, C. J. Romero, and F. H. Yocum. 1979. Development of treatment and control technology for refractory petrochemical wastes. U.S. EPA -Robert S. Kerr Envir. Research Laboratory, Ada, Oklahoma. EPA-600/2-79-080.
12. DeAngelo, D. J. and A. R. Wilhelmi. 1983. Wet air oxidation of spent caustic liquors. *Chemical Engineering Progress*, 79(3):68.
13. del Pino, M. P. and W. E. Zirk. 1982. Temperature effects on biological treatment of petrochemical wastewaters. *Environmental Progress*, 1(2):104.
14. Dennis, R. A. 1979. Water reuse in chemical plants. *Industrial Water Engineering*, 16(4):30.
15. Dold, P. L., et al. 1985. An equalization control strategy for activated sludge process control. *Water Science Technology* (G.B.) 17:p 221.
16. Eckenfelder, W. W. et al. 1985. Wastewater Treatment. *Chemical Engineering*, 92(9):60.
17. Engineering-Science, Inc. 1971. Preliminary investigational requirements - petrochemical and refinery waste treatment facilities. U.S. EPA, Water Quality Office, Project 12020 EID.
18. Environmental Protection Agency. 1973. Process Design Manual for Carbon Adsorption, Office of Technology Transfer, Washington, D.C., October 1973.

19. Figueroa, L. O. 1971. Water pollution control at Phillips Puerto Rico petrochemical complex. In: *Water 1970,* Chemical Engineering Progress Symposium Series, volume 67, number 107, p. 377.
20. Fisher, J. A., J. C. Houious, G. W. Kumke, and R. A. Conway. 1971. Pilot demonstration of basic designs for anaerobic treatment of petrochemical wastes. *Chemical Engineering Progress Symposium Series,* volume 67, number 107, p. 485.
21. Ford, D. L., J. M. Eller, and E. F. Gloyna. 1971. Analytical parameters of petrochemical and refinery wastewaters. *Journ. Water Poll. Control Fed.*, 43(8):1712.
22. Ford, D. L., G. C. Patterson, and J. M. Eller. 1973. Pollution Control in a new petrochemical complex. *Envir. Sci. and Technology,* 7(10):906.
23. Ford, D. L. and L. F. Tischler. 1974. Biological treatment best practicable control technology for treatment of refinery and petrochemical wastewaters. American Chemical Society, Division of Petroleum Chemistry, 19(3):520.
24. Fox, C. R. 1975. Remove and Recover Phenol. *Hydrocarbon Processing.* 54(7):109.
25. Gloyna, E. F. and D. L. Ford. 1970. The characteristics and pollutional problems associated with petrochemical wastes. Federal Water Pollution Control Administration. Contract number 14-12-461.
26. Gloyna, E. F. and D. L. Ford. 1979. Basis for design of industrial wastewater treatment plants. *Journal Water Poll. Control Fed.*, 51(11):2577.
27. Giusti, D. M., R. A. Conway, and C. T. Lawson. 1974. Activated carbon adsorption of petrochemicals. *Journ. Water Poll. Control Fed.*, 46(5):947.
28. Hedley, W. H., S. M. Menta, C. M. Moscowitz, R. B. Reznik, G. A. Richardson, and D. L. Zanders. 1975. *Potential Pollutants from Petrochemical Processes.* Technomic Publishing Co., Inc., Westport, Connecticut, USA.
29. Herzburn, P. A., et al. 1985. Biological treatment of hazardous waste in sequencing batch reactors. *Journal of Water Pollution Control Federation,* 57: p 1163.
30. Heukelekian, H. and M. Rand. 1955. Biochemical oxygen demand of pure organic compounds. *Sewage and Industrial Wastes,* 27, 1040.
31. Holiday, A. D. 1982. Conserving and reusing water. *Chemical Engineering,* April 19, 1982, p. 118.
32. Houious, J. C., R. A. Conway, and C. W. Ganze. 1973. Anaerobic lagoon pretreatment of petrochemical wastes. *Journal Water Pollution Control Fed.*, 45(1):71.
33. Humphrey, W. J., E. R. Witt, and J. F. Malina, Jr. 1979. Biological treatment of high strength petrochemical wastewaters. U.S. EPA. Robert S. Kerr Envir. Research Laboratory, Ada, Oklahoma. EPA-600/2-79-172.
34. Kempen, D. 1982. Manage water and energy in petrochemical plants. *Hydrocarbon Processing,* 61(8):109.
35. Khan, S. R., et al. 1985. Oxidation of 2-chloro-phenol using ozone and ultraviolet radiation. *Environ. Prog.* 4: p.229.

36. Kosarek, L. J. 1979. Water reclamation and reruse in the power, petrochemical processing and mining industries. *Proceedings of the AWWA Research Foundation, et al., Water Reuse Symposium.* Volume 1, p. 421.
37. Kulperger, R. J. 1972. Company treats its own, other wastes. *Water and Wastes Engineering,*. 9(11):F18.
38. Legan, R. W. 1982. Ultraviolet light takes on CPI role. *Chemical Engineering,* 89(2):95.
39. Mandt, M. G. 1985. The innovative technology of sequencing batch reactors. *Pollution Engineering,* 17(7):26.
40. Matthews, J. E. 1978. Treatment of petroleum refinery, petrochemical and combined industrial-municipal wastewaters with activated carbon. EPA. Robert S. Kerr Environ. Research Lab., Ada, Oklahoma. EPA-600/2-778-200.
41. McManus, A. M. C., et al. 1985. Granular Activated Carbon Removal of Priority Pollutants in a Combined Municipal/Industrial Wastewater. *Proceedings of the 39th Industrial Waste Conference, 1984, Purdue Univ.* Ann Arbor Science Publishers, Inc., Ann Arbor, MI.
42. Medley, D. R. and E. L. Stover. 1983. Effects of ozone on the biodegradability of biorefractory pollutants. *J. Water Pollut. Control Fed.*, volume 55, p. 489.
43. Meinck, F., H. Stoof, and H. Kohlschu"tter. 1968. *Industrie-Abwasser,* Gustav Fisher Verlag, Stuttgart.
44. Mencher, S. K. 1967. Minimizing waste in the petrochemical industry. *Chemical Engineering Progress,* 63(10):80.
45. Nel, L. H., J. DeHaast and T. J. Britz. 1984. Anaerobic digestion of a petrochemical effluent using an upflow anaerobic sludge blanket reactor. *Biotechnology Letters,* 6(11):741-746.
46. Nesaratnam, S. T. and F. H. Ghobrial. 1985. Biological treatment of mixed industrial and sanitary wastewaters. *Conserv. Recycl.*, 8(1-2):135-142.
47. Nijst, S. J. 1978. Treating aqueous effluents of the petrochemical industry. *Environmental Science and Technology,* 12(6):652.
48. *Oil and Gas Journal.* 1963. Here's how petrochemical companies dispose of wastes. Nov. 4, 1963.
49. Price, K. S., G. T. Waggy, and R. Conway. 1974. Brine shrimps bioassays and seawater BOD of petrochemicals. *Journ. Water Pollut. Control Fed.*, vol. 46, p. 63.
50. Pritter, P. 1976. Determination of biological degradability of organic substances. *Water Research,* vol. 10, p. 231.
51. Quartulli, O. J. 1975. Stop wastes: reuse process condensate. *Hydrocarbon Processing,* 54(10):94.
52. Rappe, G. 1985. Waste treatment process is keyed to mile-long reactor. *Chemical Engineering,* 92(8):44.
53. Rickles, R. N. 1965. Water recovery and pollution abatement. *Chemical Engineering,* September 27, 1965.
54. Rollins, R. M., et al. 1983. PACT/Wet air regeneration of an organic chemical waste. *Proceedings 37th Industrial Waste Conf., 1982, Purdue Univ.* Ann Arbor Sci. Publication, Inc., Ann Arbor, Michigan.

55. Rozich, A. F. and A. F. Gaudy. 1985. Response of phenol-acclimated activated sludge process to quantative shock loading. *Journal of Water Pollution Control Federation*, 57: p. 795.
56. Rucker, J. E. and R. W. Oeben. 1970. Wastewater control facilities in a petrochemical plant. *Chemical Engineering Progress*, 66(11):63.
57. Soderberg, R. W. and R. E. Bockrath. 1985. Treatability of diverse wastestreams in the PACT activated carbon-biological process. Proceedings of the 39th Industrial Waste Conference, 1984. Ann Arbor Science Publishers, Inc., Ann Arbor, MI. p. 121.
58. Somerville, H. J. 1985. Physiological aspects of biotreatment of petrochemical wastes. *Conserv. Recyc.*, 8(1-2)73-83.
59. Stover, E. L., et al. 1983. Ozone assisted biological treatment of industrial wastewaters containing biorefractory compounds. *Ozone Science and Engineering*, vol. 4, p. 177.
60. Tavlarides, L. L. 1983. Process modifications towards minimization of environmental pollutants in the chemical processing industry. U.S. EPA. EPA-600/2-83-120. NTIS #PB 84-133347.
61. Tavlarides, L. L. 1985. Process Modifications for Industrial Pollution Source Reduction. Lewis Publishers, Inc., Chelsea, Michigan.
62. Tchobanoglous, G. and E. D. Schroeder. 1985. Water Quality: Characteristics, Modeling, Modification. Addison-Wesley Publishing Company, Reading, Massachusetts.
63. Thibault, G. T. and T. G. Zitrides. 1979. Biomass engineering of petrochemical and refining wastewaters. In: *Mid Atlantic Industrial Waste Conference Proceedings*, volume. 11, p. 230.
64. Vinson, J. A. 1972. An improved method for the removal of organic pollutants from water. American Chemical Society, Division of Water, Air and Waste Chemistry, 12(2):8.
65. Willenbrink, R. 1973. Wastewater reuse and inplant treatment. In: *The petroleum/petrochemical industry and the ecological challenge.* American Institute of Chemical Engineers, New York, New York.
66. Wise, H. E., Jr. and P. D. Fahrenthold. 1981. Predicting priority pollutants from petrochemical processes. *Environ. Sci. and Techn.*, 15(11):1292.
67. Zeitoun, M. A., C. A. Roorda, and G. R. Powers. 1979. Total recycle systems for petrochemical waste brines containing refractory contaminants. U.S. EPA. Robert S. Kerr Environ. Laboratory, Ada, Oklahoma. EPA-600/2-79-021.

CHAPTER 4

SOLID WASTES MANAGEMENT

Introduction

Solid wastes in the petrochemical industry may occur as actual solids such as waste plastics, paper or metal; as semi-solids such as tars and resins and as suspended and dissolved solids such as waste polymers and inorganic salts. These materials may be characterized as combustible or noncombustible, organic or inorganic, inert or biodegradable, dry or mixed with either aqueous or non-aqueous liquids.

The solid wastes generated by the petrochemical industry may be stored, handled and disposed of by many different methods and combination of methods. The method or combination of methods used is dependent on existing conditions. Factors to be considered when designing solid waste processing facilities include: 1) characteristics of the wastes (volume, weight, density, ease of handling, rate of production, toxicity, biodegradability, combustibility, etc., 2) potential value of salvaged material for recycle into the same process or into new or different processes at the plant or at other plant facilities, 3) adaptability of the disposal method to the waste in question, and 4) availability of land and expected future land use patterns. Several types of solid wastes and disposal methods which have been used in the petrochemical industry are presented in Table 4.1 (Makela and Malina, 1972).

Almost every existing petrochemical manufacturing plant has some form of solid waste handling or disposal facilities on the plant premises. In a recent survey of the petrochemical industry it was observed that 90% of the solid wastes generated at petrochemical processing plants was disposed of on the plant premises (Mayhew, 1983). This means approximately 10% of these wastes are disposed of by the use of offsite or commercial disposal facilities.

Types of Solid Wastes

Solid wastes generated during the petrochemical manufacturing process include water treatment sludges, cafeteria and lunchroom wastes, plant trash, incinerator residues, plastics, metals, waste catalysts, organic chemicals, inorganic chemicals, and wastewater treatment solids. A brief discussion of each type of waste is necessary to understand the problems associated with petrochemical solid waste disposal.

Water treatment facilities may be found at many petrochemical processing plants. Solids composed of silt, sand and lime, or alum based flocculant material are

Table 4-1. Petrochemical Solid Waste Survey (Makela and Malina, 1972)

Disposal Methods Used	TYPE OF WASTE: Water Treatment Sludges	Cafeteria and Lunchroom	Plant Trash	Ashes, Flyash, Incinerator Residue	Plastic	Ferrous and Non-Ferrous Metals	Catalysts	Organic Chemicals	Inorganic Chemicals	Wastewater Treatment Sludges, Filter Cakes, and Viscous Solids
Land Disposal										
Lagoon	x						x	x	x	x
Spread on Land	x			x			x	x	x	x
Sanitary Landfill	x	x	x	x	x	x	x			x
Dumps	x	x	x	x	x	x	x		x	x
Incineration										
Stationary Hearth Furnace		x	x		x					
Multiple Hearth Furnace		x	x		x					x
Rotary Kiln		x	x		x			x	x	x
Open Pit			x		x					
Liquid Burner								x	x	
Fluidized Bed Reactor								x	x	x
Salvage and Recycle				x	x	x	x		x	
Chemical Treatment								x	x	
Biological Treatment								x		
Ocean Disposal										
Bulk Dumping								x	x	
Sealed Container Dumping								x	x	

produced during the water treatment process and require treatment and/or disposal. Cafeteria and lunch room wastes consist of food waste and paper and plastic products used in the cafeteria operation as packaging material.

Plant trash is the general term used to describe all the miscellaneous wastes which may be found at the manufacturing facility. Trash is often classified as combustible or non-combustible. Combustible material would include paper, fiberboard containers, packaging material, miscellaneous plastic and rubber products, and waste wood. Non-combustible material would include metal scraps, glass, pottery, floor sweepings, solids from storm sewers, and construction and demolition debris.

Ashes and incineration residue wastes include residues from the incineration of trash, sludges and other wastes, and residues from plant heating, steam production, water heating, and power generation facilities. The majority of the metal scraps generated in the petrochemical industry are produced during the demolition and/or construction of process equipment. Most scrap metal that is generated is bulky and of ferrous composition. This material may be contaminated by exposure to toxic substances generated during the production process.

The plastics encountered in the petrochemical industry are primarily polyethylene, polypropylene, polystyrene, and polyvinvylchloride. The reported physical form of polymer wastes encountered in the plastics industry are presented in Table 4.2 (Marynowski, 1972). Plastic particles may range in size from powders and pellets to chunks weighing more than 100 pounds. Waste plastics are generated by off-specification production, spills, product contamination, cleanout, emergency dumps,and miscellaneous other sources during plastics manufacturing (Marynowski, 1972). Plastics are generally biologically inert substances.

Table 4-2. Forms of Polymer Wastes (Marynowski, 1972)

	Average Percent of Each Form	
	Primary Resin Producers	Processors and Fabricators
Pellets	18	14
Chopped or shredded	0	3
Dust or powder	23	3
Random large (>100 lb)	10	28
Random small (<100 lb)	14	17
Other, off-specification product and contaminated product	35	35
	100	100

Spent catalysts may be liquid, semi-solid, or solid. Catalysts possess a wide range of chemical characteristics. These catalysts may possess toxic qualities and thus would require special handling. A wide variety of other organic and inorganic chemicals may enter the wastestream as a result of production processes. A large portion of these substances become part of wastewater flow streams and must be separated from these liquid wastestreams prior to treatment and/or disposal. Gloyna and Ford (1970) report that the majority of all solids found in petrochemical wastes are present as dissolved solids in liquid waste streams.

During wastewater treatment processes, suspended and/or dissolved solids are separated from wastestreams by physical-chemical unit processes producing sludges which must be disposed. Biological unit operations used in wastewater treatment processes such as activated sludge, trickling filters, extended aeration, wastewater stabilization ponds, and anaerobic digestion also produce sludges which require disposal. Wastewater treatment sludges may contain a wide variety of organic and inorganic components depending on the production process which produces the wastestream being treated.

Disposal Techniques

Solid and semi-solid waste materials generated in the petrochemical industry may be disposed of by several techniques including: salvaging and reclamation, open dump burning, no-burning dump, landfill, land farming, lagooning, incineration, and ocean dumping.

In salvaging and reclamation operations, waste materials are collected and segregated for reclamation and reuse. Salvaged materials may be reused for the original purpose or an entirely different purpose, within the same plant or outside the plant. Materials such as scrap metal, wood, spent catalyst, spent acids and caustics, contaminated oils and other hydrocarbons, plastics and polymers, rubbers and carbon black have been recovered and reused in salvage operations in the petrochemical industry (Makela and Malina, 1972).

Open dump burning is normally an unacceptable disposal alternative and may be illegal in many areas. Combustible materials are transported to an isolated location and burned in this disposal technique. The residue may or may not receive further treatment. This technique is simple, has low time and labor requirements, and has low capital and operating costs; however, this technique produces undesirable health and safety hazards and results in the production of air pollutants.

No-burn dumping involves the dumping of waste material on the ground or into pits. This technique is primarily used for non-combustible materials. This method requires large areas of land which may be rendered unsaleable for future development. Disposal by this technique also produces the potential for groundwater and surface water contamination.

Sanitary landfills are areas where wastes are buried in a controlled manner to minimize the deleterious effects on public health and environmental quality. In a sanitary landfill, refuse is confined to a small area and covered with a layer of earth each day, or more frequently, if necessary. Sanitary landfills provide the most economic environmentally acceptable method for the disposal of most non-toxic solid and semi-solid wastes generated at petrochemical processing plants (Makela and Malina, 1972).

In addition to providing the most economic environmentally acceptable disposal method for most solid wastes generated in a petrochemical manufacturing plant, sanitary landfills provide other advantages. First, a low degree of technical expertise is required to operate a landfill. Another advantage is the ease and simplicity of the operation. Disadvantages include the land requirements. Sanitary landfills require more land than other land disposal techniques. Soil and hydrogeological conditions must also be favorable.

Sanitary landfills must be constructed in areas where water will not leach through the disposal site and contaminate surface or groundwater supplies. To insure protection of water supplies, a monitoring system is desirable. This monitoring system may include drain systems around the landfill area and groundwater monitoring wells. Pritchard et al. (1984) described a three component monitoring system designed to protect water quality at an ethylene glycol plant. Landfilling practices are described in detail by Tchobanoglous (1977). Many regulatory agencies are beginning to express concern about landfill operations and careful planning must be conducted before choosing this alternative.

Another solid waste disposal technique which has been used in the petrochemical industry is "land-farming". In land-farming waste materials are spread in a thin layer over a relatively large area of land. The wastes may then be worked into the soil or left with no further treatment. This method is usually used for semi-solid materials or solids which have been mixed with liquids. The liquid portion of the waste is allowed to evaporate or percolate into the soil. The remaining solids are degraded by soil microorganisms. Land-farming has been shown to be an adequate disposal technique for petrochemical wastes of a predominately paraffinic nature (Kincannon, 1972; Francke and Clark, 1974; Dotson et al. 1970). Cihonski et al. (1978) report that land-farming may also be an adequate method for disposing of sludges containing aromatic species.

Ganze and Teller (1977) describe the operation of a land-farming, or land-spreading facility which was developed to dispose of sludge from a petrochemical processing complex wastewater treatment plant which treats 87,000 m^3/day (23 mgd). Both primary and digested waste biological sludges are disposed of on 20,000 m^2 (approximately 5 acre) plots in a 610,000 m^2 (150-acre) land-spreading area. The primary sludges disposed in this area included oily silts from API separators, chemical precipitates which are produced by pH adjustment prior to biological wastewater treatment, and solids transported from the petrochemical complex by rainfall and runoff. Sludge was pumped onto each plot to a total depth of 300-460 mm (12-18 in). The sludge is allowed to settle. Carriage water is then decanted and pumped back to the wastewater treatment plant for treatment. The sludge layer is allowed to air dry after which it is tilled into the soil.

Soil TOC and COD showed significant increases after the first sludge application. On the same plot, after subsequent applications, little change in these parameters was noted. Heavy metals concentrations appeared to be the limiting factor for determining the useful life of this land-spreading operation. Lead, nickel, manganese, copper, chromium, and arsenic concentrations increased with each sludge application. Sufficient data to estimate plot life was not obtained. Operating and maintenance costs were calculated to be US $7.75 per ton of solids disposed (1977 dollars). It was concluded that landspreading could be an effective low cost sludge disposal alternative when: 1) reasonably priced land was available, 2) climatic conditions permit, i.e., annual evaporation rate is equal to or greater than rainfall; and 3) soil types and geology preclude groundwater contamination (Ganze and Teller, 1977).

Lagooning is another technique used in the petrochemical industry. In lagooning, solid, semi-solid and liquid wastes are dumped into ponds or pits. Liquid may be discharged and possibly receive further treatment or the liquid may be retained and allowed to evaporate. Organic solids and liquids retained in the pond may be degraded biologically depending on their nature. Low construction and maintenance costs and negligible operation costs are associated with lagooning; however, lagoons used for the disposal of solid wastes generated in petrochemical production are often highly odoriferous and unsightly and have a high potential for ground and surface water pollution. Also, many of the solid and semi-solid wastes generated in petrochemical production will not degrade significantly in lagoons and will require some other method of ultimate disposal.

Incineration is a controlled combustion process for burning solid, liquid or gaseous combustible waste to gases and a residue containing little or no combustible material. Several incineration processes are used in the petrochemical industry including stationary hearth incinerators, multiple hearth furnaces, dual chamber incinerators, rotary kilns, fluidized bed reactors, open pit incinerators, and the liquid burner. The type of incinerator used in a particular application is dependent on the characteristics of the waste. Important physical and chemical characteristics which should be considered when selecting an incinerator type include material state (solid, semi-solid, liquid, gas), ease of handling, moisture content, energy value, combustion temperature, reactivity, combustion products, and ash content.

Chemical Engineering (1970) describes the operation of a dual chamber incinerator which was designed to combust the wastes generated at a petrochemical plant. The incinerator received approximately 5,670 kg/day (12,500 lb/day) of solid wastes from the petrochemical operation, a highly acidic and cokelike material with a high carbon content and low-ash content, and 1,130 kg/day (2,500 lb/day) of miscellaneous plant trash which was treated in an 8 hour work shift. The incinerator was made of two chambers, one for vaporization, the other for vapor combustion. The temperature of the first chamber was between 200 and 400°C (390 and 750°F) while the temperature of the second (combustion) chamber was approximately 1,200°C (2,200°F). Combustion of the waste left less than 5% of the original material as ash which was removed and landfilled every two to three weeks.

The dual chamber unit and the necessary waste collection facilities described in Chemical Engineering (1970) cost over US $100,000. The cost of fuel-gas to operate the incinerator was estimated to be US $5,000/year, electricity costs were estimated to be approximately US $1,000/year. One man working 40 hours/week was required to operate the facility.

Ocean dumping of solid wastes may consist of piping to near-shore waters, bulk dumping in off-shore waters, and deep sea dumping of containerized wastes. If a processing plant is located close to the sea, aqueous slurries with low organic concentrations can be pumped to near-shore waters. A tide or current which will disperse the waste material is essential. Bulk wastes such as filter cakes, sludges, and slurries may be barged to off-shore waters where they are dumped. These wastes are normally dumped 30 to 500 km (20 to 300 miles) off-shore in 300 to 500 fathoms of water by discharging the wastes through a pipe at depths of 3 to 6 m (10 to 20 ft) (Makela and Malina, 1972). Toxic liquids and sludges have been placed in containers and dumped at sea.

Smith (1971) reported that approximately 730,000 metric tons (800,000 tons) of

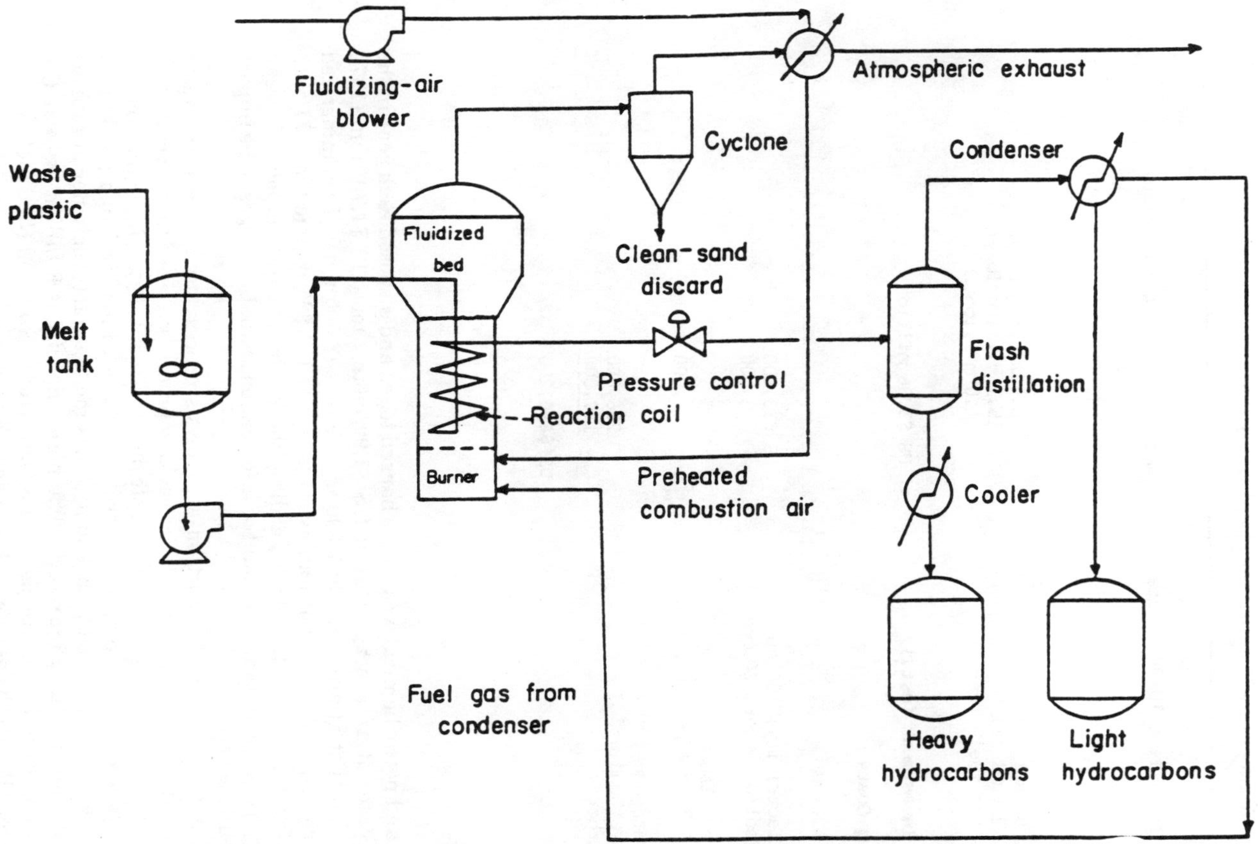

Figure 4-1. Pyrolysis Process (Bhatia and Rossi, 1982). The pyrolysis reactor is made up of three tubular coils (2 in dia.): two are used during manufacture, while the third is decoked by oxidation.

Table 4-3. Pyrolysis Process Economic Estimates (Bhatia and Rossi, 1982)

Atactic-Polypropylene Conversion Plant Economics	
Basis: 25 million lb/yr of atactic polymer (800 h/yr of operation)	
Products	
No. 6 fuel oil	6.9 million lb/yr
No. 2 fuel oil	15.1 million lb/yr
Gaseous fuels (net)	1.0 million lb/yr
Capital Investment (est.)	US $3.06 million
Operating Costs	
Utilities	
Electricity, 100 hp	US $33,600
Cooling water, 60 gpm	3,000
Direct labor (1/3 of a person per shift)	26,000
Maintenance,	
Overhead and G & A	286,050
Other	147,000
	US $496,250
Payback	2.4 yr

refinery and petrochemical wastes are barged to sea and disposed each year in the United States at an average cost of US $1.90/metric ton (US $1.70/ ton) (1970 dollars), and that approximately 7,300 metric tons (8,000 tons) of containerized wastes are disposed at sea with an average cost of US $26/metric ton ($24/ton). Little information is available regarding the environmental impact of ocean dumping; however, the small amount of information available suggests the impact may be quite severe.

Bhatia and Rossi (1982) report on a pyrolysis process used to convert waste polymers to fuel oils. This noncatalytic cracking process was used to recover almost 94% of the available fuel value from a waste plastic stream of atactic polypropylene. In the process a tubular, plug-flow pyrolysis reactor, immersed in a gas-heated fluidized bed of sand, is used to thermally crack molten atactic polymer to gaseous and liquid components (Figure 4.1). The plant capacity is 8 billion kg/year (17 billion pound/year). The atactic polymer is converted to No. 6 and No. 2 fuel oils and gaseous fuels. Research has shown that under the proper operating conditions other petrochemical wastes may be converted to gasoline additives, lighter fluid, spot remover, solvents, and petrochemical feedstocks. An estimate of the costs associated with building a conversion plant for atactic polypropylene is presented in Table 4.3 (Bhatia and Rossi, 1982).

Sludges produced during pollution control activities must also be disposed. Sludge treatment and disposal has been well studied. U.S. EPA (1979) and Newton (1985) discuss sludge treatment and disposal, including transportation, thickening, stabilization, disinfection, conditioning, dewatering, thermal processes and ultimate disposal. The techniques for estimating sludge management costs are outlined in U.S. EPA (1985).

References

1. Bhatia, J. and R. A. Rossi. 1982. Pyrolysis process converts waste polymers to fuel oils. *Chemical Engineering* 82(20):58.
2. Chemical Engineering. 1970. Dual-chamber Incinerator Burns Up Problem Wastes. *Chemical Engineering.* April 6, 1970, p. 68.
3. Cihonski, J. L., B. M. Wyman, M. P. Hughes. 1978. Disposal of petrochemical sludges by microbial decomposition. In: *American Chemical Society, Divison of Environmental Chemistry, Preprints of papers* 81(1):379.
4. Corey, R. C. 1969. *Principales and Practices of Incineration.* Wiley-Interscience, New York, NY.
5. Francke, H. C. and F. E. Clark. 1974. *Disposal of oily wastes by microbial assimilation.* U.S. Atomic Energy Commission, Y-1937.
6. Ganze, C. and J. P. Teller. 1977. Disposal of sludges by land spreading. American Society of Civil Engineers, *Journal of the Environmental Engineering Division.* 103(12):1134.
7. Gloyna, E. F. and D. L. Ford. 1970. Petrochemical Effluents Treatment Detailed. *FWPCA Water Pollution Control Research Series 12020.*
8. Kincannon, C. B. 1972. *Oily waste disposal by soil cultivation process.* EPA-R2-72-110.
9. Makela, R. G. and J. F. Malina, Jr. 1972. *Solid wastes in the petrochemical industry.* Center for Research in Water Resources. University of Texas at Austin. Austin, Texas.
10. Mayhew, J. J. 1983. Effect of RCRA on the Chemical Industry. In: *Environment and Solid Wastes.* Butterworth Publisher, Woburn, MA.
11. Marynowski, C. W. 1972. *Disposal of polymer solid wastes by primary polymer producers and plastics fabricators.* Solid Wastes Office, U.S. Environmental Protection Agency, SW-34C.
12. Newton, J. 1985. Special Report: Sludge treatment and disposal processes. *Pollution Engineering,* 17(12):16.
13. Pritchard, R. B., H. D. Sharma, and M. J. Kollmeyer. 1984. Three component monitoring system guards Canadian petrochemical plant's groundwater. *Oil and Gas Journal* 82(18)53.

14. Smith, D. and R. P. Brown. 1971. Ocean disposal of barge-delivered liquid and solid wastes from U.S. Coastal Cities. Solid Wastes Office, U.S. Environmental Protection Agency. SW-19c.
15. Tchobanoglous, G., H. Theisen, and R. Eliassen. 1977. *Solid wastes: Engineering principles and management issues.* McGraw-Hill Book Company, New York.
16. U.S. EPA. 1979. *Process Design Manual for Sludge Treatment and Disposal.* U.S. EPA Municipal Environmental Research Laboratory, Cincinnati, OH. EPA-625/1-79-011.
17. U.S. EPA. 1985. *Estimating Sludge Management Costs.* Water Engineering Research Laboratory, Cincinnati, OH. EPA-625/6-85/010.

CHAPTER 5

DISPOSAL OF HAZARDOUS WASTES

Hazardous wastes pose a potential hazard to the health of humans or other living organisms because the wastes are lethal, nondegradable, persistent in nature, can be biologically magnified or otherwise cause detrimental cumulative effects (EPA, 1974). The U.S. Environmental Protection Agency characterizes a waste as hazardous if it possesses any one of the following four characteristics: (1) ignitability, (2) corrosivity, (3) reactivity, or (4) toxicity (Section 4, U.S. Code of Federal Regulations, Part 261).

An ignitable waste is any waste which will present a fire hazard during routine management. A corrosive waste is any waste which is able to deteriorate standard containers or to dissolve toxic components of other wastes. A reactive waste is any waste which has a tendency to become chemically unstable under normal management conditions, which will react violently when mixed with water, or which will generate toxic gases. A toxic waste is any waste which will pose a substantial hazard or potential hazard to human health. Based on these definitions and the previous discussions of air and water pollution and solid wastes generated during petrochemical manufacturing many of these wastestreams may be characterized as hazardous.

There are hundreds of documented cases of damage to life and the environment which have resulted from the improper management of hazardous wastes. Hazardous wastes have been found to contaminate ground water supplies, rivers, lakes and other surface waters, polluted the air, caused fires and explosions, caused serious illness by contaminating foodstuffs and by direct contact (Kiang and Metry, 1982). These wastes are frequently bioaccumulated, very persistent in the environment and often toxic at very low concentrations. The source of the vast majority of these cases may be traced back to some part of the petrochemical industry.

Harm to human health and the environment caused by past mismanagement of hazardous waste has led to increased public concern about hazardous waste management. Proper management means more than just careful disposal. A range of management options must be considered. In order of priority the desired options for managing hazardous wastes are (Kiang and Metry, 1982):

(1) minimizing the amount of waste generated by modifying the industrial process involved
(2) transfer the waste to another industry which may use the waste
(3) reprocess the waste to recover materials and energy
(4) separate hazardous and nonhazardous materials
(5) subject the waste to some process which will render the waste nonhazardous
(6) dispose of the waste in a secure landfill

There are an estimated two million recognized chemical compounds with more than 60,000 chemical substances in past or present commercial use. Approximately 600 to 700 new chemicals, mostly synthetic organics produced from petrochemicals, are introduced each year; but published reports of animal testing have been issued for only about 15,000. Some of these substances may possess carcinogenic, mutagenic, and teratogenic effects which may be extended in time, perhaps for 10, 20 , or 30 years, to the point where direct relationships with morbidity and mortality are difficult to conclusively prove (Salvato, 1982). For these reasons the proper management of hazardous wastes in the petrochemical industry is very important.

The survey conducted by Hedley et al. (1975) of 190 petrochemical production processes (listed in Table 3.1) is very useful in identifying hazardous wastestreams that originate from these processes. A survey of the hazardous wastestreams from 24 organic chemical, pesticide, and explosives manufacturing plants was conducted by Process Research, Inc. (1977). In this survey 16 organic chemical manufacturing, 5 pesticide manufacturing, and 3 explosives manufacturing industries (all of which use petrochemicals as feedstocks) were surveyed and the major hazardous wastestreams from each industry were identified. Table 5.1 contains a list of these industries and the identified hazardous wastestreams.

The U.S. Environmental Protection Agency has identified 129 toxic hazardous wastes which have become known as "priority pollutants". Wise and Fahrenthold (1981) in a study of petrochemical processes identified these hazardous "priority pollutants" in many petrochemical process wastewaters. Table 3.3 contains a list of critical precursor and generic process combinations that result in the generation of these "priority pollutants" at signficant levels. A list of wastestreams from organic chemical manufacturing processes containing "priority pollutants" may be found in Table 1.2 (Wise and Fahrenthold, 1977). A list of plastics/synthetic fibers manufacturing processes which contain these "priority pollutants" may be found in Table 5.2.

Currently, most of the process wastes from the petrochemical manufacturing industry are ultimately destined for land disposal or in some cases incineration. Alternative treatment processes were investigated by Process Research, Inc. (1977). A summary of these processes in shown in Tables C. 1 and C. 2 in Appendix C. There are many alternative treatment processes available which may be classified as physical, chemical, or biological processes. The costs associated with these treatment alternatives varies widely, and the costs of all options must be compared before selecting a treatment process. The alternative process considered most desirable was economically compared to the sanitary landfill, chemical landfill, and incineration alternatives. A list of the results of this comparison may be found in Table C.3 in Appendix C. It can be seen from the data contained in this table that these alternatives are economically favorable in many cases, and in some cases actually produce an economic benefit as in the case of the alternative process for perchloroethylene manufacture. In this case distillation of the waste reduces the waste volume by 90 percent and recovers hexachlorobutadien, producing a net savings of US $378 per kg of waste processed.

Transferring a hazardous waste to another industry has received increasing attention. This process may take place in a materials exchange to handle, treat and physically exchange wastes or an information exchange. Such a clearinghouse leaves generators and purchasers to negotiate directly. The first information exchange was established in the Netherlands in 1972. Since then the idea has spread through Europe and is growing in the United States (Kiang and Metry, 1982).

With shortages of raw materials and more restrictive disposal regulations, recovery has become a more attractive alternative. Many wastes contain valuable substances which can be extracted from concentrated wastestreams more economically than processing from virgin materials.

Incineration has proved to be a viable method of destroying organic wastes without posing a threat to the environment. Chlorine or bromine-containing compounds have been destroyed successfully in cement kilns and special incinerators aboard ships at sea (Kiang and Metry, 1982).

For further study on hazardous waste management in the petrochemical manufacturing industry several textbooks are currently available (Kiang and Metry, 1982; Lindgren, 1983; Metry, 1980; and Pojasek, 1979 a, b; 1980, 1982). Published proceedings of recent industrial waste treatment conferences and symposia are also good sources of information on hazardous waste treatment, management and disposal (see Boardman, 1986 and Bell, 1985, 1986 for example).

Tragedies such as those that have occured in Love Canal, New York and Kentucky's "Valley of the Drums" have focused attention on what can happen when hazardous wastes are improperly managed. Technologies exist for environmentally sound management, but these have not been widely used because they appear to be costly and because often there is no legal requirement for their use. In many cases it is impossible to assign monetary values to the long-term damage to health and the environment that has resulted from improper management of hazardous wastes. But the astronomical costs of cleaning up damage caused by poor disposal practices alone is reason enough to justify the cost of proper environmental controls. In this case an ounce of prevention is a sound investment.

Table 5-1. Hazardous Wastesteams Identified in Some Petrochemical Manufacturing Processes (Process Research, Inc., 1977)

Product and Typical Plant Size	Hazardous Wastestream Components	Waste Generation KKg*/yr
Perchloroethylene 39,000 KKg/yr	Hexachlorobutadiene Chlorobenzenes Chloroethanes Chlorobutadiene Tars	12,000
Nitrobenezene 20,000 KKg/yr	Crude Nitrated Aromatics	50
Chloromethane 50,000 KKg/yr	Hexachlorobenzene Hexachlorobutadiene Tars	300
Epichlorohydrin 75,000 KKg/yr	Epichlorohydrin Dichlorohydrin Chloroethers Trichloropropane Tars	4,000
Toluene Diisocyanate 27,500 KKg/yr	Polyurethane Ferric Chloride Isocyanates Tars	358
Vinyl Chloride Monomer 136,000 KKg/yr	1, 2 Dichloroethane 1, 1, 2 Trichloroethane 1, 1, 1, 2 Tetrachloroethane Tars	1,400
Methyl Methacrylate 55,000 KKg/yr	Hydroquinone Polymeric Residues	4,730
Acrylonitrile 80,000 KKg/yr	Acrylonitrile Higher Nitriles	160
Maleic Anhydride 11,000 KKg/yr	Maleic Anhydride Fumaric Acid Chromogenic Compounds Tars	333
Lead Alkyls 60,000 KKg/yr	Lead	30,000
Zthanolamines 14,000 KKg/yr	Triethanolamine Tars	1,120
Furfural 35,000 KKg/yr	Sulfuric Acid Tars & Polymers	19,600
Furfural 35,000 KKg/yr	Fines & Particulates From Stripped Hulls	350
Fluorocarbon 80,000 KKg/yr	Antimony Pantachloride Carbon Tetrachloride Trichlorofluoromethane Organics	18

Table 5-1. (continued)

Product and Typical Plant Size	Hazardous Wastestream Components	Waste Generation KKg*/yr
Chlorotoluene 15,000 KKg/yr	Benzylchloride Benzotrichloride	15
Chlorobenzene 32,000 KKg/yr	Polychlorinated Aromatic Resinous Material	1,400
Atrazines 20,000 KKg/yr	Water Sodium Chloride Insoluble Residues Caustic Cyanuric Acid	224,600
Trifluralin 10,000 KKg/yr	Spent Carbon Fluoroaromatics Intermediates and Solvents	1,150
Malathion 14,000 KKg/yr	Filter Aid Toluene Insoluble Residues Dimethyl Dithiophosphoric Acid	1,816
Malathion 14,000 KKg/yr	Malathion Toluene Impurities Sodium Hydroxide	14,350 (W) 350 (D)
Parathion 20,000 KKg/yr	Diethylthiophosphoric Acid	2,300
Explosives 93,000 KKg/yr	Activated Carbon Nitrobodies (Any organic nitrated byproduct)	330 (W) 200 (D)
Explosives 30,000 KKg/yr	Redwater (Waste from purification of crude TNT)	15,000
Explosives 125,000 KKg/yr	Waste Explosives	250

*1 KKg = 1 Metric Ton (MT)
(W) Wet Basis
(D) Dry Basis

Table 5-2. Plastics/Synthetic Fibers Effluents with Concentrations Greater than 0.5 ppm of Priority Pollutants (Wise and Fahrenthold, 1977)

Product	Monomer(s)	Associated priority pollutants
ABS resins	Acrylonitrile Styrene Polybutadiene	Acrylonitrile Aromatics
Acrylic Fibers	Acrylonitrile Comonomer (variable): Vinyl Chloride	Acrylonitrile Chlorinated C2's
Acrylic resins (Latex)	Acrylonitrile Acrylate ester Methylmethacrylate	Acrylonitrile Acrolein
Acrylic resins	Methylmethacrylate	Cyanide
Alkyd resins	Glycerin Isophthalic acid Phthalic anhydride	Acrolein Aromatics Polyaromatics
Cellulose acetate	Diketene (acetylating agent)	Isophorone
Epoxy resins	Bisphenol A Epichlorohydrin	Phenol Chlorinated C3's Aromatics
Petroleum hydrocarbon resins	Dicyclopentadiene	Aromatics
Phenolic resins	Phenol Formaldehyde	Phenol Aromatics
Polycarbonates	Bisphenol A Phosgene	(Not investigated) Predicted: phenol Chloroaromatics Halomethanes
Polyester	Terephthalic acid/ dimethylterephthalate Ethylene glycol	Phenol Aromatics
HD polyethylene resin	Ethylene	Aromatics
Polypropylene resin	Propylene	Aromatics
Polystyrene	Styrene	Aromatics
Polyvinvy chloride resin	Vinyl chloride	Chlorinated C2's
SAN resin	Styrene Acrylonitrile	Aromatics Acrylonitrile
Styrene-Butadiene resin (Latex)	Styrene (50%) polybutadiene	Aromatics
Unsaturated polyester resin	Maleic anhydride Phthalic anhydride Propylene glycol (Styrene-added lated)	Phenol Aromatics

References

1. Bell, J. M., Editor. 1985. *Proceedings of the 39th Industrial Waste Conference, 1984.* Butterworth Publishers, Stoneham, MA.
2. Bell, J. M., Editor. 1986. *Proceedings of the 40th Industrial Waste Conference, 1985.* Butterworth Publishers, Stoneham, MA.
3. Boardman, G. P. 1986. *Toxic and Hazardous Waste.* Technomic Publishing Company, Inc., Lancaster, PA.
4. EPA, Office of Solid Waste Management Programs. 1974. Report to Congress: Disposal of Hazardous Wastes. U.S. Environmental Protection Agency, Publication SW-115. Washington, D.C.
5. Kiang, Y.H. and A. A. Metry. 1982. *Hazardous Waste Processing Technology.* Ann Arbor Science Publishers, Inc., Ann Arbor, Michigan.
6. Lindgren, G. F. 1983. *Guide to Managing Industrial Hazardous Waste.* Butterworths Publishers, Woburn, MA.
7. Metry, A. A. 1980. *The Handbook of Hazardous Waste Management.* Technomic Publishing Company, Westport, CT.
8. Pojasek, R. B. 1979a. *Toxic and Hazardous Waste Disposal,* Volume 1, Processes for Stabilization/Solidification. Ann Arbor Science Publishers, Inc., Ann Arbor, Michigan.
9. Pojasek, R. B. 1979b. *Toxic and Hazardous Waste Disposal,* Volume 2, Options for Stabilization/Solidification. Ann Arbor Science Publishers, Inc., Ann Arbor, Michigan.
10. Pojasek, R.B. 1980, *Toxic and Hazardous Waste Disposal,* Volume 3, Impact of Legislation and Implementation on Disposal Management Practices. Ann Arbor Science Publishers, Inc., Ann Arbor, Michigan.
11. Pojasek, R.B. 1982. *Toxic and Hazardous Waste Disposal,* Volume 4, New and Promising Ultimate Disposal Options. Ann Arbor Science Publishers, Inc., Ann Arbor, Michigan.
12. Process Research, Inc. 1977. Alternatives for Hazardous Waste Management in the Organic Chemical, Pesticides and Explosives Industries. U.S. Environmental Protection Agency, Hazardous Waste Management Division, Washington, D.C., EPA/530/SW-151c. NTIS #PB 278059.
13. Salvato, J. A. 1982. *Environmental Engineering and Sanitation.* Wiley-Interscience, New York, New York.

CHAPTER 6

PETROCHEMICALS INDUSTRY IN DEVELOPING COUNTRIES

Worldwide consumption of petrochemical products is expected to increase by approximately 5 percent per year between 1980 and 2000. Petrochemical consumption in the developing countries is expected to grow at an even higher rate during this same period. Petrochemical demands are expected to grow at an annual rate of 7.8 percent in Latin America, 8 percent in North Africa and the Middle East, 6.2 percent in South Asia, and 9 percent in Southeast Asia (UNIDO, 1983 a). Many of these developing countries are rich in hydrocarbons and other raw materials necessary for petrochemical production. The availability of the necessary raw materials, an inexpensive labor force, and an increased demand for petrochemical products is expected to lead to the development of petrochemical production capabilities within these developing countries. The number of new plants needed to meet the predicted increased demand for petrochemical products in developing countries is listed in Table 6.1 (UNIDO, 1983 b).

The information presented in previous chapters of this book has shown that petrochemical production will result in the generation of water and air pollutants, solid and hazardous wastes. An increase in petrochemical production could, therefore, have a significant impact on public health and environmental quality. The information contained in this book has also shown that the technology to control these potential pollutant emissions currently exists.

To avoid the adverse effects on public health and environmental quality of this increased petrochemical production, adequate pollution control regulations must be promulgated. Such regulations would require the evaluation of possible environmental impacts and public health effects of the construction and operation of petrochemical production facilities, and require measures to mitigate adverse effects.

To assess possible environmental impacts, a survey must be conducted of the existing environmental conditions at the proposed plant site. A survey of the wastes generated at a plant should also be conducted. The survey should include a characterization of the volume of wastes generated, the rate of flow of the wastes, and the physical, chemical, and biological characteristics of the generated wastes.

The need for environmental regulation of petrochemical production has been recognized in some developing countries. Tewari et al. (1980) described pollution

Table 6-1. Increase in Demand and Estimated New Investment for Major Petrochemicals in Developing Countries by 1990 (in thousand metric tons) (UNIDO, 1983 b)

	Increased 1990 demand over 1984 capacity	Capacity and number of new units	Investment cost of new units $ million
OLEFINS	8 030	10 000 (23)	10 350
AROMATICS	2 680	*	*
PLASTICS	11 440	10 850 (124)	14 201
POLYESTER SYN. FIBERS	990	975 (27)	1 727
SBR SYNTHETIC RUBBER	720	725 (23)	1 623
METHANOL	240	250 (1)	90
AMMONIA	15 248	16 900 (56)	9 090
TOTAL			37 081

* included in Olefins plants

control efforts in the petrochemical industry in India. Industry in India is concentrated in a few limited areas such as Baroda, Bombay, Calcutta, and Kanpur. The Water/Air Pollution (Prevention and Control) Act, 1974/78 and the constitution of central and state boards for water pollution control contain regulations designed to protect the environment from industrial pollutants in India. Tewari et al. (1980) noted that proper waste management in India has not only helped abate pollution problems but has also improved economic viability of various industries when the most efficient and economical pollution control facilities were used.

The government of the Taiwan Province of China has developed a special industrial estate in southern Taiwan Province in which wastewater treatment is one of the most important considerations (Peng et al. 1978). The Linyuan Industrial Estate is a 380 hectare industrial estate designed for petrochemical industries. The Estate is located on the banks of the Kaoping River and will contain more than 30 industries. Primary and intermediate petrochemical products were expected to be produced at this complex. A list of some of the water pollution standards in the Taiwan Province, which had to be met by this project, are found in Table 6.2 (Peng et al. 1978). Several alternative treatment processes were evaluated including primary treatment with an ocean outfall, secondary treatment with a short-

Table 6-2. Water Standards for Various Uses in Taiwan, China (Peng et al., 1978)

I. Fresh Water - Rivers, Lakes and Ponds
Conventional Constituents

Class	Best Usage	pH	Standard of Water Biological Oxygen Demand (5 day, 20°C) mg/L	Dissolved Oxygen mg/L	Coliform Bacteria Median No/100 mL (MPN)	Suspended Solids mg/L
AA	Public water supply (I) bathing, or any lesser use	6.5–8.5	1	6.5	50	
A	Public water supply (II), fishing (I), or any lesser use	6.0–9.0	2	5.5	5,000	25
B	Public water supply (III), fishing (II), industrial water supply (I), or any lesser use	6.0–9.0	4	4.5	10,000	40
C	Irrigation and Industrial water supply (II) or any lesser use	6.0–9.0		2		100
D	Environmental protection	6.0–9.0		2		No floating

Organics and Metals

Cyanide mg/L	Organic Phosphate mg/L	Cd mg/L	Pb mg/L	Cr + 6 mg/L	As mg/L	Total Hg mg/L	Se mg/L	Phenol mg/L
0.01	None detectable	0.01	0.1	0.05	0.1	0.005	0.05	0.001

*The Above Water Standards are Based on the River Flow of Yearly Duration at 75%.
Water Supply (1) : Source of drinking water after disinfection.
Water Supply (II) : Source of drinking water after conventional water treatment process.
Water Supply (III): Source of drinking water after additional treatment other than conventional process.

distance marine outfall, and tertiary treatment with river discharge. Secondary treatment with the improved Kraus Process of air aeration activated sludge followed by a marine outfall was selected (Figure 6.1).

Pollution control was also a concern at the Zulia El Tablazo Petrochemical Complex in Venezuela (Ford et al. 1973). The construction of the giant Zulia El Tablazo Petrochemical Complex on the shore of Lake Maracaibo at El Tablazo in the State of Zulia represented a joint venture between the Venezuelan government and private corporations. The capital investment exceeded US $1.2 billion in 1979. It was recognized by the government that this facility represented a potential source of pollution to an ecologically sensitive body of water that was already receiving various levels of pollution input. To avoid further damage to Lake

Table 6-2. (continued)

II. Marine Waters
Conventional Constituents

			Standard of Water				
Class	Best Usage	pH	Biological Oxygen Demand (5 day, 20°C) mg/L	Dissolved Oxygen mg/L	Coliform Bacteria Median No/100 mL (MPN)	Suspended Solids mg/L
A	Fishing (I), bathing, or any lesser use	7.5–8.5	2	6	1,000	2
B	Fishing (II), industrial water supply (II), or any lesser use	7.5–8.5	3	5		3
C	Environmental protection	7.0–8.5	8	2		8

Fishing (I) : Fresh water for silver carp & grass carp, marine water for striped mullet & sea weeds.
Fishing (II) : Fresh water for carps & shellfish, marine water for milk fish.

Industrial Water Supply (I) : Industrial water for processing use.
Industrial Water Supply (II) : Industrial water for cooling use.

Organics and Metals

The Restrictions Over the Content of Cyanide, Organic Phosphate, and Heavy Metals are the Same as Those Over Fresh Water.

Maracaibo, the government initiated a pollution control program through the Instituto Venezolano de Petroquimica (IVP). The program was designed to contain and treat wastewaters discharged from the petrochemical production facilities to a level that would not damage the estuary.

The project was planned and implemented under the manifestation of the Venezuelan governmental policy for industrial development with environmental control. At the time of initiating the project, there were no specific discharge standards established by the government. It was the responsibility of IVP to establish acceptable effluent standards and to predict the impact of the discharge of the treated wastewater. Guidelines prepared for similar situations in the United States of America were selected as guides and used to develop the treatment program. The environmental impact assessment (EIA) was limited to the wastewater treatment associated with the petrochemical complex. Air pollution and solid wastes disposal, excluding sludge disposal for the wastewater treatment process, were not considered in the assessment.

The economic and social impacts were assessed by the Venezuelian government, and a decision to proceed with construction was made. Economic and social factors, other than the protection of the uses of the estuary, were not a factor in assessing the need for the wastewater treatment facility.

The first step in the assessment of wastewater treatment needs was to conduct a base line survey to determine the quality of the water in the estuary before

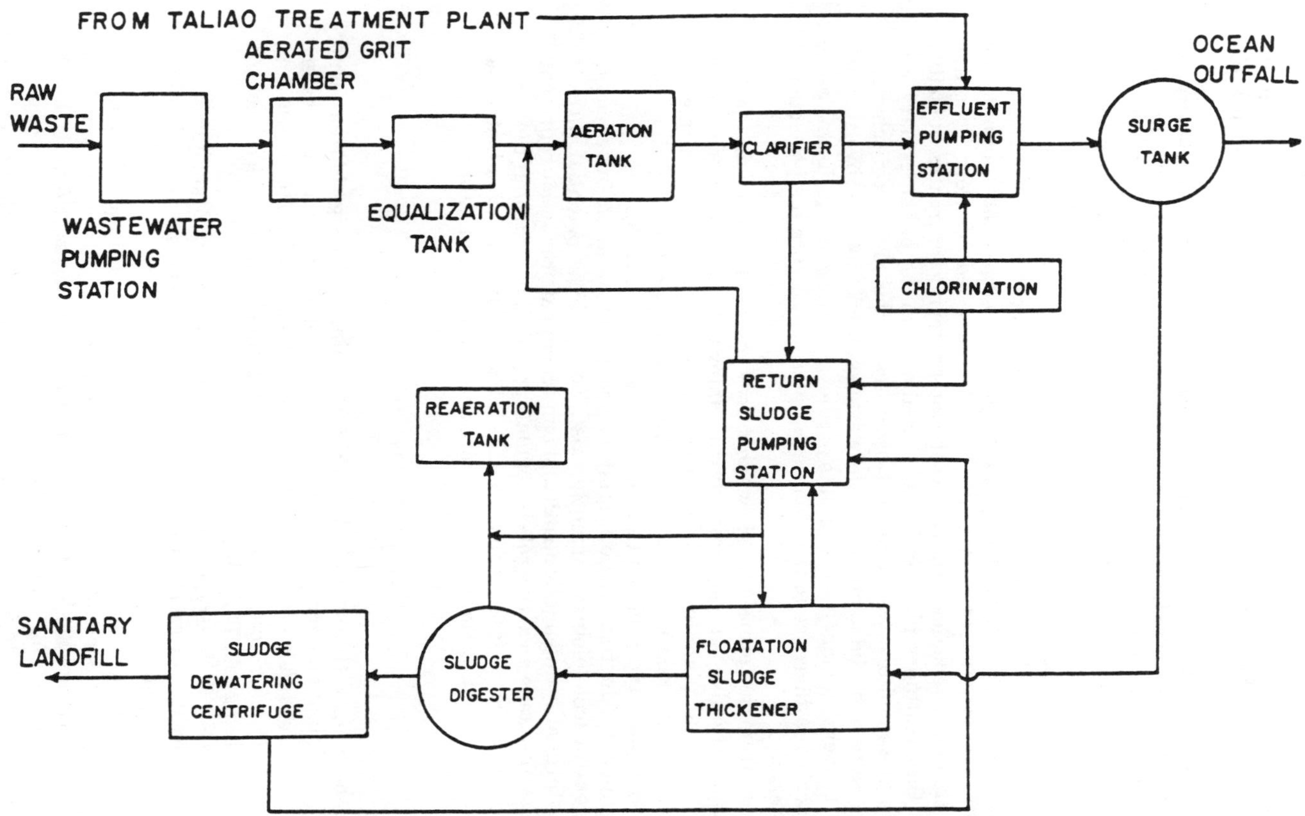

Figure 6-1. Schematic Flow Diagram of Linyuan Industrial Wastewater Treatment Plant (Peng et al., 1978)

constructing the plant. Water samples were collected at various locations and depths during the various seasons of the year and analyzed for chemical and biological content. Results of dispersion studies were used to develop a one dimensional model to predict the movement of water in the estuary.

Various types of wastewater treatment processes were evaluated on a laboratory scale, and the results of these tests were used to select a treatment process capable of producing an effluent quality acceptable for discharge to the estuary.

Samples of wastewaters from similar petrochemical processes were collected for analysis and composited in proportion to the volume of wastewater expected from the El Tablazo Complex. This composited wastewater was used in the treatability studies and served as a basis for design of the wastewater treatment facility.

A detailed monitoring scheme was prepared and incorporated into the operating plan for the wastewater treatment facility. Sampling schedules were devised to determine the characteristics of the raw wastewater entering the treatment facility as well as the treatment plant effluent. Periodic sampling of the estuary was included in the monitoring scheme. All processes at the complex discharging wastewater were required to prove that the wastewater was not toxic by using fish bioassays. Dispersion models were used to predict the impact of the discharges to Lake Maracaibo. Adherence to the monitoring plan will assure the protection of Lake Maracaibo.

Experience from this project has shown that answers to potential environmental problems can be resolved only by properly combining governmental initiative and support with sound planning and engineering. Governmental commitment to environmental protection and control has produced a thorough investigation, a comprehensive plan, and proficient engineering (Ford et al. 1973).

Petrochemical production can be a significant source of pollution, but adequate, economic control technologies currently exist. To avoid the potential threat to environmental quality and public health that petrochemical production represents, governments must take an active role in regulating pollution control. If the proper steps are taken, the benefits that come from the introduction of a new industry may be realized while avoiding damage to environmental quality and public health.

References

1. Peng, K. H., C. Y Huang, C. F. Wu. 1978. Planning and design of treatment plant and marine outfall of petrochemical industrial wastewater. In: *International Conference on Water Pollution Control in Developing Countries*. Volume 2. Asian Institute of Technology, Bangkok, Thailand.
2. Tewari, R. N., K. Rudrappa, K. L. L. Narasimhan. 1980. Management of wastes in petroleum industry. *Indian Chemical Manufacturer*, Volume 29.
3. UNIDO. 1983 a. World Demand for Petrochemical Products and the Emergence of New Producers from the Hydrocarbon Rich Developing Countries. Sectorial Studies Series, No. 9. UNIDO, Division for Industrial Studies.
4. UNIDO. 1983 b. Opportunities for co-operation among the developing countries for the establishment of the petrochemical industry. Sectoral Working Paper series, No. 1, UNIDO, Division for Industrial Studies.

CHAPTER 7

ENERGY CONSIDERATION IN POLLUTION CONTROL

Energy use at pollution control facilities must be considered with respect to three different areas of growing concern, the direct cost of the energy used, the environmental effects of pollution generated directly and indirectly as a result of energy use, and depletion of important nonrenewable resources. Each of these concerns should play a role in the selection of processes used in pollution control facilities.

Rapidly changing energy prices are forcing pollution control facilities operators to give serious consideration to the energy requirements of pollution control. Research has shown that energy costs will become the predominant factor in the selection of some pollution control facility alternatives (Middlebrooks et al. 1981). The wide variety of petrochemical manufacturing processes and thus the wide range of pollution control devices used at these facilities makes it difficult to give much detail about energy use for pollution control in the petrochemical manufacturing industry as a whole; however, some information is available. Table 7.1 is an example of the tables of information available for many processes used in the petrochemical industry (Air Products and Chemicals, Inc., 1974 a, b; 1975 a, b, c, d, e, f, g). These tables contain information about air pollution control alternatives for several different petrochemical manufacturing processes and the energy cost for each air pollution control alternative. Knowing the basis of these cost estimates (which is also found in each table i.e., electricity costs were assumed to be US \$0.01/kw-hr) power requirements may also be computed for each alternative process.

An analysis of the data contained in Tables 7.1 and companion tables showed that energy costs may be as much as 99 percent of the total annual operating cost (minus depreciation and interest on capital) of an air pollution control process. Some air pollution control alternatives may result in a net energy production, as in the use of a boiler house vent gas burner on absorber vent gas in the manufacture of formaldehyde with the silver catalyst process. The impact that energy use may have on the operating cost of different pollution control alternatives can be estimated from these tables presented in the reports by Air Products and Chemicals, Inc. (1974 a, b; 1975 a, b, c, d, e, f, g).

Table 7-1. Cost Effectiveness of Alternative Emission Control Devices Used in the Manufacture of Phthalic Anhydride from Ortho-Xylene (based on 130 million lbs/yr phthalic anhydride production) (Air Products and Chemicals, Inc., 1975 e)

Stream	Main Process Vent Gas					Waste Products
Type of Emission Control Device	Water Scrubber +	Incineration	Direct Incineration(g)	Incineration +	Waste Heat Boiler	Direct Incineration
Number of Units	2	1	2	2	2	1
Capacity of each Unit - %	50	100	50	50	50	100
Feed Gas						
Total Flow - Lbs/Hr	536,962	10,460 (b)	536,962	536,962		5,792
SCFM	119,300		119,300	119,300		
Composition - Ton/Ton PAN						
Hydrocarbons						
Particulates (Inc PAN,MAN & Org Acids)	0.0756(a)	0.1221	0.0692	0.0692		0.0557
NO_x						
SO_x	0.0047		0.0047	0.0047		
Carbon Monoxide	0.1507		0.1507	0.1507		
Gaseous Effluent						
Total Flow - Lbs/Hr	544,005	39,235	540,052	545,767		19,497
SCFM	122,100	9,700	120,200	122,350		4,900
Composition - Ton/Ton PAN						
Hydrocarbons						
Particulates (Inc PAN,MAN & Org Acids)	0.0036	0.0009	0.0036	0.0036		0.0004
NO_x		0.0002	0.0006	0.0012		0.0001
SO_x	0.0047		0.0047	0.0047		
Carbon Monoxide	0.1507	0.0026	0.0076	0.0076		0.0025
Emissions Control Efficiency (d)						
CCR		98	95	95		97
SERR	86	99	92	92		99
SE	96 (organics)					

PAN = Phthalic Anhydride
MAN = Maleic Anhydride

Table 7-1 (continued)

Stream	Main Process Vent Gas				Waste Products
Type of Emission Control Device	Water Scrubber +	Incineration	Direct Incineration[g]	Incineration + Waste Heat Boiler	Direct Incineration
Investment - US $					
Purchased Cost	275,000	120,000	576,000	625,000	85,000
Installation	825,000	730,000	285,000	625,000	65,000
Total Capital [c]	1,100,000	850,000	860,000	1,250,000	150,000
Operating Cost - US $/Yr					
Depreciation (10 years)	110,000	35,000	86,000	125,000	15,000
Interest on Capital (6%)	66,000	21,000	51,600	75,000	9,000
Maintenance	55,000 (5%)	35,000 (10%)	34,400 (4%)	50,000 (4%)	7,500 (5%)
Labor - US $4.85/Hr	6,500	5,000	5,000	20,000	3,000
Utilities and Chemicals					
Power - US $0.01/KWH	25,000	5,000	19,800		1,000
Fuel - US $0.40/million BTU		55,500	198,300	562,300	12,800
Process Water - US $0.10/mil gal	1,100				
Boiler Feed Water - US $0.30/mil gal				34,200	
Total Utilities and Chemicals	26,100	60,500	218,100	596,500	13,800
Total Operating Cost:	263,600	156,500	395,100	866,500	48,300
Steam Production-US $0.59/LBs (450 PSIG, 750°F)				(465,000)[f]	
Net Annual Cost - US $/Yr	420,100		395,100	401,500	48,300

(a) Includes 0.0064 T/T of organic material contained in separate liquid reject stream from product fractionation system ejector.
(b) Liquid rejected from scrubber system plus light and heavy ends removed in product fractionation.
(c) It is possible that future fuel cost will be considerably higher than figure used in this comparison.
(d) Emission control efficiencies are defined by the equations given below.

$$\text{CCR} = \frac{\text{pounds of } O_2 \text{ that react with pollutants to feed device}}{\text{pounds of } O_2 \text{ that theoretically could react with these pollutants}} \times 100$$

$$\text{SERR} = \frac{\text{weighted pollutants in} - \text{weighted pollutants out}}{\text{weighted pollutants in}} \times 100$$

$$\text{SE} = \frac{\text{specific pollutant in} - \text{specific pollutant out}}{\text{specific pollutant in}} \times 100$$

(e) Developed from 1970-1971 cost figures provided by PAN manufacturer with 10-15 percent added for escalation to 1973 costs.
(f) Shown at fuel plus BFW cost since this steam only replaces operating cost of stand-by boilers.
(g) With feed preheat.

Very little information is available concerning energy requirements of wastewater treatment processes when used in the petrochemical manufacturing industry; however, many of the processes used for wastewater treatment in the petrochemical industry are similar or identical to the processes used in municipal wastewater treatment. The energy data compiled for municipal treatment facilities may be used to estimate the energy requirements of similar processes in the petrochemical industry, taking into account the differences in wastewater characteristics and economies of scale.

Wesner et al. (1978) presented a detailed analysis of energy requirements by unit operations and unit processes employed in municipal wastewater treatment. The results of the Wesner et al. (1978) study were presented in graphical form, with accompanying tables outlining the design considerations employed in developing the graphs. Energy requirements were presented in terms of the design flow rate of the treatment system in most cases, but when a wide choice of loading rates was applicable, the graphs were presented in terms of surface area or of flow rate applied to the component of the system. Using these more detailed energy usage data will be helpful in estimating the energy requirements of petrochemical wastewater treatment facilities.

Middlebrooks et al. (1981) presented analyses of the energy requirements of small wastewater treatment systems, including advanced physical-chemical treatment processes which may be necessary when treating petrochemical wastestreams which contain complex synthetic organics. It was concluded that increasing energy costs were assuming a greater proportion of the annual operating costs of wastewater treatment facilities of all sizes, and could become the predominant factorin selecting cost-effective treatment alternatives. It was observed that low energy consuming treatment systems were generally easier to operate and maintain than energy-intensive systems, making low energy consuming systems even more attractive. Simple biological processes were found to require much less energy than mechanical and physical-chemical systems when they are applicable (Middlebrooks et al. 1981).

Culp (1978) presented an analysis of alternatives for future wastewater treatment at an advanced treatment facility that illustrates the sensitivity of energy costs. Energy use was not considered in the original design of the advanced wastewater treatment facility in the late 1960's. The energy required for alternative processes is compared to the energy required by the original design in Table 7.2 (Culp, 1978; Culp and Culp, 1971). It was anticipated that the final effluent from the flood irrigation alternative would be at least equal in quality to the effluent from the original physical chemical process.

McMillan et al. (1981), Garber et al. (1975), Hagan and Roberts (1976), and Mills and Tchobanoglous (1974) all present similar discussions of the energy requirements of wastewater treatment processes. Burris (1981), Prindle et al. (1983), and Jacobs (1977) all discuss methods to manage energy usage and energy cost in wastewater treatment.

In addition to the cost for pollution control energy use, consideration also must be given to the environmental effects of pollution generated directly and indirectly as a result of energy use. The generation of power produces pollution. The amount of pollution produced as a direct result of energy production is a function of the power generation process used, the fuel used, the amount of power produced and the pollution control facilities at the point of generation. Because power generation

Table 7-2. Energy Requirement of a 2.8 x 10^4 m^3/d (7.5 mgd), Advanced Wastewater Treatment System (Culp, 1978; Culp and Culp, 1971)

Alternative	Total energy[a] (electricity and fuel expressed as equivalent M kWh/yr)[b]
Original system complete secondary treatment, AWT system, effluent export to Indian Creek Reservoir (storage reservoir)	64,500
1978 alternatives	
Continue secondary, nitrification, effluent export to Indian Creek Reservoir	39,400
Continue secondary, nitrogen removal (ion exchange) effluent export to Indian Creek Reservoir	40,244
Continue secondary on-site, flood irrigation land treatment in Carson River Basin	25,000

[a]Does not include secondary energy requirements for chemical manufacture

[b] 1 kWh = 3.6 x 10^6J

produces pollution, the use of energy indirectly results in the production of pollution. The processing of the fuels used in energy production also produces significant impacts on environmental quality, and this impact can be considered to be pollution generated indirectly from the use of energy. Since it is the goal of pollution control facilities to produce the least environmental impact within cost constraints, it is necessary to consider this "indirect" generation of pollution when choosing between pollution control alternatives.

Finally, as has been previously shown, some pollution control alternatives may require large amounts of energy while other alternatives result in a net energy savings. When energy use is required by a pollution control device, serious consideration should be given to a basic resource management problem. Should energy, which is often available only in the form of a nonrenewable resource, be used to conserve water quality, which is a naturally renewable resource? These decisions are very difficult and will require a greater knowledge of the energy requirements of pollution control facilities and available energy resources than is currently available.

References

1. Air Products and Chemicals, Inc. 1974a. Engineering and cost study of air pollution control for the petrochemical industry. Volume I: carbon black manufacture by the furnace process. U.S. EPA, EPA-450/3-73-006-a. Research Triangle Park, NC. NTIS#PB-238 324.
2. Air Products and Chemicals, Inc. 1974b. Engineering and cost study of air pollution control for the petrochemical industry. VolumeIII: ethylene dichloride manufacture by oxychlorination. U.S. EPA,EPA—450/3-73-006-c. NTIS#PB-240 492.
3. Air Products and Chemicals, Inc. 1975a. Engineering and cost study of air pollution control for the petrochemical industry. VolumeII: Acrylonitrile Manufacture. U.S. EPA, EPA-450/3-73-006-b. NTIS#PB-240 986.
4. Air Products and Chemicals, Inc. 1975b. Engineering and cost study of air pollution control for the petrochemical industry. Volume IV: formaldehyde manufacture with the silver catalyst process. U.S. EPA, EPA-450/3-73-006-d. NTIS#PB-242 118.
5. Air Products and Chemicals, Inc. 1975c. Engineering and cost study of air pollution control for the petrochemical industry. Volume V: formaldehyde manufacture with the mixed oxide catalyst process. U.S. EPA, EPA-450/3-73-006-e. NTIS#PB-242 547.
6. Air Products and Chemicals, Inc. 1975d. Engineering and cost study of air pollution control for the petrochemical industry. Volume VI: ethylene oxide manufacture by direct oxidation of ethylene. U.S. EPA, EPA-450/3-73-006-f. NTIS#PB-244 116.
7. Air Products and Chemicals, Inc. 1975e. Engineering and cost study of air pollution control for the petrochemical industry. Volume VII: phthalic anhydride manufacture from ortho-xylene. U.S. EPA, EPA-450/3-73-006-g. NTIS#PB-245 277.
8. Air Products and Chemicals, Inc. 1975f. Engineering and cost study of air pollution control for the petrochemical industry. Volume VIII: vinyl chloride manufacture by the balanced process. U.S. EPA, EPA-450/3-73-006-h. NTIS#PB-242 247.
9. Air Products and Chemicals, Inc. 1975g. Engineering and cost study of air pollution control for the petrochemical industry. VolumeIX: polyvinyl chloride manufacture. U.S. EPA, EPA-450/3-73-006-i. NTIS#PB-247 705.
10. Burris, B. E. 1981. Energy conservation for existing wastewater treatment plants. *Water Pollution Control Federation Journal*, 53(5):536.
11. Culp, R. L. and G. L. Culp. 1971. *Advanced Wastewater Treatment.* Van Nostrand Reinhold Company, New York, NY.
12. Culp, G. L. 1978. Alternatives for Wastewater Treatment at South Tahoe, California. Paper presented at the 51st Annual Conference of the Water Pollution Control Federation, Anaheim, California. October 1978.
13. Garber, W. F., G. T. Ohara, and S. K. Raksit. 1975. Energy-Wastewater Treatment and Solids Disposal. *American Society of Civil Engineers, Journal of the Environmental Engineering Division,* June 1975, p. 319.

14. Hagan, R. M. and E. B. Roberts. 1976. Energy Requirements for Wastewater Treatment. *Water and Sewage Works*, 123(12):51.

15. Jacobs, A. 1977. Reduction and Recovery: Keys to Energy Self-Sufficiency. *Water and Sewage Works*, Reference Number R-24-R-37.

16. McMillan, H. H., R. R. Rimkus, and F. C. Neil. 1981. Metro Chicago's study of energy alternatives for wastewater treatment. *Water Pollution Control Federation Journal*, 53(2):155.

17. Middlebrooks, E. J., C. H. Middlebrooks, S. C. Reed. 1981. Energy requirements for small wastewater treatment systems. *Journal of the Water Pollution Control Federation*, 53(7):1172.

18. Mills, R. A. and G. Tchobanoglous. 1974. Energy Consumption in Wastewater Treatment. In: *Energy, Agriculture and Waste Management*, W. J. Jewell (Ed.), Ann Arbor Science Publishers, Ann Arbor, Michigan.

19. Prindle, W.M., D.A. Buell, and L.J. Scully, 1983. A management approach to energy cost control in wastewater utilities. *Water Pollution Control Federation Journal*, 55(10):1239.

20. Wesner, G. M. et al. 1978. Energy Conservation in Municipal Wastewater Treatment. EPA 430/9-77-01. U.S. Environmental Protection Agency, Office of Water Program Operations, Washington, D.C.

CHAPTER 8

SUMMARY AND CONCLUSIONS

Petrochemicals include a wide variety of compounds which are listed in several international standard industrial classifications including: industrial gases, cyclic intermediates, dyes, organic pigments and crudes, organic chemicals, inorganic chemicals derived from petroleum plastic materials, insecticides and agricultural fertilizers. Primary petrochemicals, produced from raw materials such as crude petroleum, natural gas, heavy fractions such as fuel oil, etc., include alkynes, olefins, paraffins, aromatics, hydrogen, hydrogen sulfide and carbon black. These first generation compounds are used as feedstocks in the synthesis of intermediate and third generation petrochemical products.

A wide variety of chemical reactions and unit processes may be included in petrochemical processes. Currently more than 500 different processing sequences are used in the petrochemical industry. This leads to a very complex waste problem.

Air Pollution Control

Air pollutants produced by petrochemical manufacturing practices include sulfur oxides, nitrogen oxides, carbon monoxide, particulate matter, odors and a wide variety of toxic and nontoxic hydrocarbons. Petrochemical plants discharge pollutants into the atmosphere that are either controlled or fugitive in nature. Controlled emissions are released through stacks and/or vents, and detailed information is available on emissions composition and rate of release. Emissions from points other than stacks and vents are considered fugitive emissions. Fugitive emissions may occur due to accidents, inadequate maintenance, poor planning, and from a range of process equipment such as valves, pumps, flanges, compressors and agitators.

Control of air pollutants emitted from controlled sources has been well studied. Many texts are available detailing the design of pollution control equipment for these sources (Crawford 1976, Stern 1968 a, b, and c, Danielson 1967, Strauss 1966 and 1972, Nonhebel 1972).

Valves, flanges and pump seals are the biggest contributors to fugitive emissions at petrochemical plants. Proper selection and maintenance of valves, flanges and pumps will reduce fugitive emissions and eliminate potential product losses which have been estimated to be over US $1800/day at a typical olefin plant. Techniques to measure fugitive emission rates from petrochemical plants have been described by Hughes et al. (1979) and Siversten (1983).

In a survey of petrochemical plants in the United States (US) the US Environmental Protection Agency determined that the manufacture of carbon black resulted in the emission of the largest mass of air pollutants with the manufacture of acrylonitrile a distant second. The mass of emissions is not the only criterion that must be considered in assessing the impact of petrochemical plant air pollutant emissions. The toxicity of emissions, odors and the persistence of the emitted compounds are some additional considerations.

The main difference between air emissions from petrochemical plants and other industrial processes is the emission of a wide variety of hydrocarbon compounds. Many of these hydrocarbons are considered toxic, and thus special precautions must be taken for their control. Hydrocarbon emission reduction systems at petrochemical plants are described by Pruessner and Broz (1977), Kenson (1979) and Mashey and McGrath (1979). Kenson (1979) presented five concepts for the design of an organic chemical emission control system:

(1) Thoroughly define the problem
(2) Define the degree of control required
(3) Weigh advantages and disadvantages of alternative control systems
(4) Evaluate total cost (operating and capital) associated with the systems
(5) Design the final system choices to optimally control the particular waste stream

Some techniques to reduce hydrocarbon emissions without emission control systems include: (1) appropriate specification, selection and maintenance of seals in valves, flanges and pumps, (2) installation of floating roof tanks to control evaporation of light hydrocarbons, (3) installation of vapor recovery lines to vents of vessels that are continually filled and emptied, (4) manifolding of purge lines used for start-ups and shutdowns to vapor recovery or flare systems, (5) venting of vacuum jet exhaust lines to vapor recovery systems, (6) shipment of products by pipeline, (7) covering waste separators, and (8) use of steam or air injection at flares.

Costs of air pollution control systems vary widely with the process and degree of control desired. The higher the removal rate required, the higher the removal cost per unit mass of pollutant removed. Many techniques which reduce air emissions produce economic benefits by reducing product loss and recovering usable compounds.

Wastewater Treatment and Disposal

Wastewater streams in the petrochemical production industry may be categorized into six source components:

(1) wastes discharged directly from production units during normal operation
(2) utility operations such as blow down from energy production and cooling systems

(3) sanitary sewage from administrative areas, locker rooms, shower and restroom facilities, and food handling areas
(4) contaminated stormwater runoff from process areas
(5) ballast water discharged from tankers during product handling
(6) miscellaneous discharges from spills, turnarounds, etc.

The most commonly used method for predicting the quality and quantity of petrochemical production wastewaters is to study each individual unit process, and relate the quantity and quality of the wastestreams produced to the production units. For example, the isopropanol stripping still and intermediate flash column used in acetone production produces approximately 2.2 pounds of Chemical Oxygen Demand (COD) per ton of acetone produced. This is a difficult task because small changes in unit process operating conditions alter the characteristics of the wastestream produced.

Gloyna and Ford (1970) conducted a survey designed to characterize petrochemical production wastes. As a part of this survey many petrochemical wastestreams were described in terms of conventional pollution parameters such as acidity, alkalinity, color, turbidity, pH, Biochemical Oxygen Demand (BOD), COD, Total Organic Carbon (TOC), solids, surface activity, taste, odor, and temperature. The characteristics of the wastestreams vary so widely it is impossible to make any generalizations. It is important to note, however, that petrochemical wastestreams may be very significant sources of many toxic substances.

The design of wastewater treatment facilities for petrochemical facilities will not be reliable unless wastewaters have been fully characterized and the performance characteristics of alternative treatment processes have been evaluated by treatability studies and pilot plant operations. Treatability studies should establish the effects of operational parameters such as hydraulic detention time, sludge age and temperature on organic removal rates, oxygen requirements, sludge production, sludge characteristics and process stability. Treatability studies can also identify wastestreams which should be treated separately to enhance process performance.

The unit processes capable of treating petrochemical manufacturing plant wastewaters are as varied as the unit processes used in the manufacturing plants themselves. Studies have shown that there are seldom cost effective alternatives to biological treatment used in conjunction with physical-chemical pretreatment and/or polishing where needed (Ford and Tishler, 1974; Nijst, 1978). Biological treatment coupled with post-filtration has been defined by the US Environmental Protection Agency as the "best practicable technology" currently available for treating petrochemical processing wastewaters.

Special attention must be given to the removal of toxic substances from petrochemical processing wastewaters. These toxic substances are frequently not removed during biological treatment and may require the use of other treatment processes such as activated carbon adsorption, chemical oxidation, steam stripping, solvent extraction, polymeric adsorption, chemical coagulation and sedimentation, wet air oxidation or pyrolysis.

The petrochemical industry lends itself to controlling pollution through process improvement rather than pollution abatement. Four alternative solutions may be developed for a pollution problem in the petrochemical industry. First, some wastes may be recovered as salable coproducts. Second, wastestreams can be recycled after some process modification for conversion to prime product or for reuse in the process as a reagent or intermediate. Third, the waste may be usable as a fuel. Fourth, and least desirable, wastes may be treated in waste treatment processes where they are converted to less harmful states and/or dispersed in quantities which may be assimilated by the environment. Many processes for wastewater treatment fitting into the first three categories are available in the petrochemical industry. Many techniques are also available for reducing the amount of water used at petrochemical plants, thus reducing the amount of wastewater to be treated.

As in air pollution control wastewater treatment costs vary significantly; however, as pointed out previously, the basic approach to pollution control will significantly effect pollution control costs. Many systems which have been designed to reduce pollution by eliminating the pollution at the source, recovering materials which have some economic benefit, or conserve water have resulted in an economic benefit rather than a cost. Burgess (1973) reported that one US petrochemical company installed 450 pollution abatement projects with a total cost of US $20,000,000 in 1971. The net annual savings from these projects was estimated to be US $6,000,000, with an annual return on investment of 30 percent.

Solid Waste Management

Solid wastes in the petrochemical industry may occur as actual solids such as waste plastics, paper or metal; as semi-solids such as tars and resins, and as suspended and dissolved solids such as waste polymers and inorganicsalts. These wastes include water treatment sludges, cafeteria and lunchroom wastes, plant trash, incinerator residues, plastics, metals, waste catalysts, organic chemicals, inorganic chemicals, and wastewater treatment solids. The materials may be characterized as combustible or non-combustible, organic or inorganic, inert or biodegradable, dry or mixed with either aqueous or nonaqueous liquids.

The solid wastes generated by the petrochemical process may be managed by many different methods which are dependent on existing conditions such as: (1) characteristics of the wastes (volume, weight, density, rate of production, toxicity, biodegradability, etc.), (2) potential value of salvaged materials, (3) adaptability of the disposal method to the waste of interest, and (4) availability of land and expected land use patterns. Almost every petrochemical plant has some form of solid waste handling and/or disposal facilities on the plant premises. A recent survey of the petrochemical industry disclosed that 90 percent of the solid wastes generated at petrochemical processing plants was disposed of on plant premises (Mayhew, 1983).

Solid and semi-solid waste materials generated by the petrochemical industry may be disposed of by several techniques including: salvaging and reclamation, open dump burning, no-burning dump, landfill, land farming, lagooning, incineration, and ocean dumping.

Salvaging and reclamation operations are environmentally acceptable operations in which waste materials are collected and segregated for reclamation and reuse. Materials such as scrap metal, wood, spent catalyst, spent acids and caustics, contaminated oils and other hydrocarbons, plastics and polymers, rubbers and carbon black have been recovered and reused in salvage operations in the petrochemical industry (Makela and Malina, 1972).

Open dump burning and no-burn dumping are normally considered unacceptable alternatives since they pose significant threats to public health and environmental quality.

Sanitary landfills provide the most economical environmentally acceptable method for the disposal of most non-toxic solid and semi-solid wastes generated at petrochemical processing plants (Makela and Malina,1972). In addition to the economic advantage, another advantage of sanitary landfills is that a low degree of technical expertise is required for operation. Soil and hydrogeological conditions must be favorable to prevent contamination of surface and groundwater supplies by water which may leach through the disposal site. Land-farming, lagooning, incineration and open-dumping may also prove to be acceptable alternatives under the proper conditions.

Hazardous Wastes Control

Many wastes generated by the petrochemical industry must be considered hazardous wastes. Hazardous wastes may be defined as any waste or combination of wastes which pose a substantial hazard or potential hazard to the health of humans or other living organisms because the wastes are lethal, nondegradable, persistent in nature, can be biologically magnified or otherwise cause detrimental cumulative effects (EPA, 1974). The US EPA characterizes a waste as hazardous if it possesses any one of the following four characteristics: (1) ignitability, (2) corrosivity, (3) reactivity, or (4) toxicity. Hazardous wastes have been identified in petrochemical wastestreams by Hedley et al. (1975), Process Research, Inc. (1977) and Wise and Fahrenthod (1977).

There are literally hundreds of documented cases of damage to life and the environment resulting from the improper management of hazardous wastes. These wastes are frequently bioaccumulated, very persistent in the environment and often toxic at very low concentrations. The source of the vast majority of these cases may be traced back to some part of the petrochemical industry.

Currently most of the process wastes from the petrochemical manufacturing industry are ultimately destined for land disposal or in some cases incineration. There are many alternative treatment processes available which may be classified as physical, chemical or biological and may be economically favorable to land disposal or incineration. These alternatives are evaluated in a report prepared by Process Research, Inc. (1977) and are specific to the individual processes.

The desired options for managing hazardous wastes, listed in order of priority are (Kiang and Metry, 1982):

(1) minimizing the amount of waste generated by process modification
(2) transfer the waste to another industry for use
(3) reprocess the waste to recover materials and energy
(4) separate hazardous and nonhazardous wastes
(5) subject the waste to some process which will render the waste non-hazardous
(6) dispose of the waste in a secure landfill

Hazardous waste management in the petrochemical industry is a very complex problem. In many cases it is impossible to assign monetary values to long-term damage to health and the environment that has resulted from improper management of hazardous wastes. The astronomical costs of cleaning up damage caused by poor disposal practices alone is reason enough to justify the cost of proper environmental controls. Several textbooks are currently available which discuss the problem of hazardous waste management (Kiang and Metry, 1982; Lindgren, 1983; Metry, 1980; and Pojasek, 1979 a, b; 1980, 1982).

Energy Use

Energy use at pollution control facilities must be considered with respect to three different areas of growing concern, the direct cost of the energy used, the environmental effects of pollution generated directly and indirectly as a result of energy use, and depletion of important nonrenewable resources. Rapidly changing energy prices are forcing pollution control facilities operators to give serious consideration to the energy requirements of pollution control. Research has shown that energy costs will become the predominant factor in the selection of some pollution control facility alternatives (Middlebrooks et al. 1981). For example, energy costs may be as much as 99 percent of the total annual operating cost of some air pollution control processes.

In addition to the cost of energy used for pollution control, consideration must be given to the environmental effects of pollution generated directly and indirectly as a result of energy use. The generation of power produces pollution. The processing of fuels used for energy production also results in the generation of environmental impacts. Since some pollution control alternatives may require large amounts of energy while other alternatives may result in a net energy savings, and since it is the goal of pollution control facilities to produce the least environmental impact within cost constraints, it is necessary to consider these costs of power generation when choosing between control alternatives.

Industry Growth

Worldwide consumption of petrochemical products is expected to increase by approximately 5 percent per year between 1980 and 2000. Petrochemical consumption in the developing countries is expected to grow at an even higher rate during this same period. Petrochemical demands are expected to grow at an annual rate of 7.8 percent in Latin America, 8 percent in North Africa and the Middle East, 6.2 percent in South Asia, and 9 percent in Southeast Asia (UNIDO, 1983). Many of these developing countries are rich in hydrocarbons and other raw materials necessary for petrochemical production. The availability of the necessary raw materials, an inexpensive labor force, and an increased demand for petrochemical products is expected to lead to the development of petrochemical production capabilities within these developing countries.

Petrochemical production will result in the generation of water and air pollutants, solid and hazardous wastes. An increase in petrochemical production could, therefore, have a significant impact on public health and environmental quality. The technology to control these potential pollutant emissions currently exists.

To avoid the adverse effects on public health and environmental quality of this increased petrochemical production, adequate pollution control regulations must be promulgated. Such regulations would require the evaluation of possible environmental impacts and public health effects of the construction and operation of petrochemical production facilities, and require measures to mitigate adverse effects.

To assess possible environmental impacts, a survey must be conducted of the existing environmental conditions at the proposed plant site. A survey of the wastes generated at a plant should also be conducted. The survey should include a characterization of the volume of wastes generated, the rate of flow of the wastes, and the physical, chemical, and biological characteristics of the generated wastes.

Petrochemical production can be a significant source of pollution, but adequate, economic control technologies currently exist. To avoid the potential threat to environmental quality and public health that petrochemical production represents, governments must take an active role in regulating pollution control. If the proper steps are taken, the benefits that come from the introduction of a new industry may be realized while avoiding damage to environmental quality and public health.

Management Philosophy

It is advantageous to consider excess materials as an additional resource to be utilized either in the form discarded or after further processing. This approach to waste processing is economically and environmentally important. If a government or ministry considers protection of the environment and maximum utilization of the base resource important, then the production management and the employees probably have an entirely different attitude toward performing this function and are more likely to take pride in producing high quality effluents and in recovering and utilizing as much of the material as possible. The importance of protecting the quality of the environment and the impact that improper handling of waste materials has on the employees' life styles and the nation as a whole must be emphasized.

Environmental protection must be stressed when management is expected to meet production quotas. Under such production systems management tends to concentrate its talent on product output, if not reminded continually of the value placed on environmental protection by the ministry and the nation. Environmental protection must be considered as a valuable natural resource in the same manner as the labor, materials, and the capital investment required to produce the basic product.

The costs for environmental protection must be paid either now or in the future. The most effective method of handling excess products is to incorporate the facilities for protecting the environment and for further processing of the excess into useful products. It is much less expensive to install such equipment initially than to convert a production process and add pollution control equipment later; moreover, it has proved cheaper to spend today's currency than an inflated one of a later date. However, it is still less expensive to add to existing systems the facilities for processing excess materials than to allow excess to be wasted as environmental pollutants; to clean these up at a future time is costly and difficult. Indeed, the damage to the environment before installing equipment to correct a situation may be impossible to rectify. It is burdensome to assess the economic losses incurred by people and industry because of delayed pollution control; however, these are real economic factors which must be considered and emphasized. The losses of health, happiness, and productivity of people owing to environmental pollution are the greatest costs of all.

Long-term economic effects of industrial pollution must not be neglected. If an industry is allowed to develop in an area without pollution control facilities, eventually the area may deteriorate to a level unacceptable to many of the residents, and they move away. Relocation of the population depletes the tax base for public services and results in a further deterioration of the local living conditions. With an added tax burden the community is forced to extract more support from the industry, resulting in higher product costs. Environmental pollution also influences maintenance costs for homes, public buildings, and thoroughfares, as well as the industrial buildings and equipment themselves.

Pollution control is a good business practice which a nation cannot afford to neglect. Maintenance of the environment is much the same as maintenance of machinery, automobiles, and other devices. If a nation does not routinely care for the environment, eventually it deteriorates. Deterioration may occur to a level that is intolerable to flora and fauna and cost the people and the government more than the industry produces. A nation must not sacrifice its customs and desirable environment to short-term economic advantage.

Some form of industrial waste treatment must be practiced if degradation of environmental quality is to be prevented. Complete treatment at the industrial site may be necessary, pretreatment prior to discharge to a public sewer may be required, or discharge to a treatment facility serving an industrial complex may provide the effluent quality needed. The degree of treatment required varies with local and national standards and the economy of by-product recovery.

References

1. Burgess, K. L. 1973. Clean up environmental problems in the petrochemical and resin industry can be profitable. In: *The Petroleum/Petro-Chemical Industry and the Ecological Challenge.* American Institute of Chemical Engineers. New York, NY.
2. Crawford, M. 1976. *Air Pollution Control Theory.* McGraw-Hill, Inc., New York, NY.
3. Danielson, J. A., ed. 1967. *Air Pollution Engineering Manual, Public Health Service Publication 999-AP-40,* National Center for Air Pollution Control, Cincinnati, OH (available from U.S. Government Printing Office, Washington, DC).
4. EPA, Office of Solid Waste Management Programs. 1974. Report to Congress: Disposal of Hazardous Wastes. U.S. Environmental Protection Agency, Publication SW-115. Washington, D.C.
5. Ford, D. L. and L. F. Tischler. 1974. Biological treatment best practicable control technology for treatment of refinery and petrochemical wastewaters. American Chemical Society, Division of Petroleum Chemistry, 19(3):520.
6. Gloyna, E. F. and D. L. Ford. 1970. The characteristics and pollutional problems associated with petrochemical wastes. Federal Water Pollution Control Administration. Contract number 14-12-461.
7. Hedley, W. H., S. M. Mehta, C. M. Moscowitz, R. B. Reznik, G. A. Richardson, D. L. Zanders. 1975. *Potential Pollutants from Petrochemical Processes.* Technomic Publications, Westport, CT.
8. Hughes, T. W., D. R. Tierney, and Z. S. Khan. 1979. Measuring fugitive emissions from petrochemical plants. *Chemical Engineering Progress.* 75(8):35-39.
9. Kenson, R. E. 1979. Engineered Systems for the Control of Toxic Chemical Emissions. Proceedings of the 72nd Annual Meeting of the Air Pollution Control Association. Paper #79-17.4.
10. Kiang, Y. H. and A. A. Metry. 1982. *Hazardous Waste Processing Technology.* Ann Arbor Science Publishers, Inc., Ann Arbor, Michigan.
11. Lindgren, G. F. 1983. *Guide to Managing Industrial Hazardous Waste.* Butterworth Publishers, Woburn, MA.
12. Makela, R. G. and J. F. Malina, Jr. 1972. *Solid wastes in the petrochemical industry.* Center for Research in Water Resources. University of Texas at Austin. Austin, Texas.
13. Mashey, J. H. and J. J. McGrath. 1979. Engineering Design of Organics Emission Control Systems. Presented at the 1979 National Petroleum Refiners Association Annual Meeting.
14. Mayhew, J. J. 1983. Effect of RCRA on the Chemical Industry. In: *Environmental and Solid Wastes,* Butterworth Publisher, Woburn, MA.
15. Metry, A. A. 1980. *The Handbook of Hazardous Waste Management.* Technomic Publishing Company, Westport, CT.
16. Middlebrooks, E. J., C. H. Middlebrooks, S. C. Reed. 1981. Energy requirements for small wastewater treatment systems. *Journal of the Water Pollution Control Federation,* 53(7):1172.

17. Nijst, S. J. 1978. Treating aqueous effluents of the petrochemical industry. *Environmental Science and Technology*, 12(6):652.

18. Nonhebel, G., ed. 1972. *Processes for Air Pollution Control*, CRC Press, Cleveland, OH.

19. Pojasek, R. B. 1979a. *Toxic and Hazardous Waste Disposal*, Volume 1, Processes for Stabilization/Solidification. Ann Arbor Science Publishers, Inc., Ann Arbor, Michigan.

20. Pojasek, R. B. 1979b. *Toxic and Hazardous Waste Disposal*, Volume 2, Options for Stabilization/Solidification. Ann Arbor Science Publishers, Inc., Ann Arbor, Michigan.

21. Pojasek, R. B. 1980. *Toxic and Hazardous Waste Disposal*, Volume 3, Impact of Legislation and Implementation on Disposal Management Practices. Ann Arbor Science Publishers, Inc., Ann Arbor, Michigan.

22. Pojasek, R. B. 1982. *Toxic and Hazardous Waste Disposal*, Volume 4, New and Promising Ultimate Disposal Options. Ann Arbor Science Publishers, Inc., Ann Arbor, Michigan.

23. Process Research, Inc. 1977. Alternatives for Hazardous Waste Management in the Organic Chemical, Pesticides and Explosives Industries. U.S. Environmental Protection Agency, Hazardous Waste Management Division, Washington, D.C., EPA/530/SW-151c. NTIS #PB 278059.

24. Pruessner, R. D. and L. B. Broz. 1977. Hydrocarbon Emission Reduction Systems. *Chemical Engineering Progress*, 73(8):69-73.

25. Siversten, B. 1983. Estimation of diffuse hydro-carbon leakages from petrochemical factories. *Journal of the Air Pollution Control Association* 33(4):323-327.

26. Stern, A. C., ed. 1968a. *Air Pollution*, Volume I. "Air Pollution and its effects". Academic Press, Inc., New York, NY.

27. Stern, A. C., ed. 1968b. *Air Pollution*, Volume II. "Analysis, Monitoring and Surveying". Academic Press, Inc., New York, NY.

28. Stern, A. C., ed. 1968c. *Air Pollution*, Volume III. "Sources of Air Pollution and their control". Academic Press, Inc., New York, NY.

29. Strauss, W. 1966. *Industrial Gas Cleaning*. Pergamon Press, London, England.

30. UNIDO. 1983. World Demand for Petrochemical Products and the Emergence of New Producers from the Hydrocarbon Rich Developing Countries. Sectorial Studies Series, No. 9. UNIDO, Division for Industrial Studies.

31. Wise, H. E., Jr. and P. D. Fahrenthold. 1981. Predicting priority pollutants from petrochemical processes. *Environ. Sci. and Techn.*, 15(11):1292.

APPENDIX A

GLOSSARY

ABSORPTION: The process by which one substance is taken into and included within another substance, as the absorption of water by soil and nutrients by plants.

ACIDITY: Quantitative capacity of aqueous solutions to react with hydroxylions. Measured by titration, with a standard solution of a base to a specified end point. Usually expressed as milligrams per liter of calcium carbonate.

ACTIVATED CARBON: Carbon "activated" by high-temperature heating with steam or carbon dioxide, producing an internal porous particle structure. Total surface area of granular activated carbon is estimated to be 1,000 m^2/gm.

ACTIVATED SLUDGE PROCESS: A biological wastewater treatment process in which a mixture of wastewater and activated sludge is agitated and aerated. The activated sludge is subsequently separated from the treated wastewater (mixed liquor) by sedimentation and wasted or returned to the process as needed.

ADSORPTION: Adhesion of an extremely thin layer of molecules (gas or liquid) to the surfaces of solids (e.g., granular activated carbons) or liquids with which they are in contact.

AERATE: To permeate or saturate a liquid with air.

AEROBIC: (a) Having molecular oxygen as a part of the environment. (b) Growing or occurring only in the presence of molecular oxygen, such as aerobic organisms.

AGGLOMERATION: A phenomenon where particles mass together.

ALKALINE: Presence of the hydroxides, carbonates, and bicarbonate of elements, such as calcium, magnesium, sodium, potassium; or of ammonia. Alkaline pH values ranges from 7.1 to 14.

ALKALINITY: Capacity of water to neutralize acids, imparted by the water's content of carbonates, bicarbonates, hydroxides, and occasionally borates, silicates, and phosphates. Expressed in milligrams per liter of equivalent calcium carbonate.

AMORTIZATION: The serial repayment of principal.

ANAEROBIC: (a) The absence of molecular oxygen. (b) Growing in the absence of molecular oxygen (such as anaerobic bacteria).

ANAEROBIC CONTACT PROCESS: An anaerobic waste treatment process in which the microorganisms responsible for waste stabilization are removed from the treated effluent stream by sedimentation or other means, and held in or returned to the process to enhance the rate of treatment.

ANAEROBIC WASTE TREATMENT: Waste stabilization brought about by the action of microorganisms in the absence of air or elemental oxygen. Usually refers to waste treatment by methane fermentation.

AQUACULTURE: The culture of fish or other aquatic life in water.

ASSIMILATIVE CAPACITY: Capacity of a natural body of water to receive (a) wastewaters, without deleterious effects; (b) toxic materials, without damage to aquatic life or humans consuming the water; and (c) BOD, within prescribed dissolved oxygen limits.

AUTOTROPHIC: Self-nourishing: denoting the green plants and those forms of bacteria that do not require organic carbon or nitrogen, but can form their own food out of inorganic salts and carbon dioxide.

BIOASSAY: Assay method using a change in biological activity as a qualitative or quantitative means of analyzing the response of biota to industrial wastes and other wastewaters. Viable organisms, such as live fish or daphnia, are used as test organisms.

BIOCHEMICAL OXYGEN DEMAND (BOD_5): The 5-day, 20°C, BOD_5 test is widely used to determine the pollutional strength of wastewater in terms of oxygen required to oxidize or convert the organic matter to a nonputrescible end product. The BOD_5 test is a bioassay procedure that measures the oxygen consumed by living organisms while utilizing the organic matter present in the wastewater under conditions as similar as possible to those that occur in nature. To make results comparable, the test has been standardized. The BOD_5 test is one of the most important in stream pollution control.

BIOLOGICAL OXIDATION: Process in which living organisms in the presence of oxygen convert the organic matter contained in wastewater into a more stable or mineral form.

BLOWDOWN: Periodic or continuous draw-off of a mixture from a system to prevent buildup of contaminants.

BOD: Biochemical Oxygen Demand

CAPITAL COSTS: The costs of the project from its beginning to the time the works are placed in operation. Included are (a) the purchase of property and rights-of-way; (b) payments for equipment and construction and for engineering and legal services; and (c) interest charges during construction.

CATALYTIC INCINERATORS: Incinerators for gaseous materials which utilize a catalyst to reduce the operation temperature.

CATION EXCHANGE: The interchange between a cation in solution and another cation on the surface of any surface-active matrial, such as clay or organic colloids.

CHEMICAL COAGULANT: Destabilization and initial aggregation of colloidal and finely divided suspended matter by the addition of floc-forming chemical.

CHEMICAL OXYGEN DEMAND (COD): The COD test is an alternative to the BOD_5 test. It is widely used and measures the quantity of oxygen required to oxidize the materials in wastewater under severe chemical and physical conditions. The major advantage of the COD test is that only a short period (3 hours) is required to conduct the test. The major disadvantage is that the test does not indicate how rapidly the biologically active material would be stabilized in natural conditions.

CHEMICAL PRECIPITATION: Separating a substance from a solution, resulting in the formation of relatively insoluble matter.

CHLORINATION: Application of chlorine to water or wastewater, generally for the purpose of disinfection, but frequently for accomplishing other biological or chemical results.

CHLORINE RESIDUAL: The total amount of chlorine (combined and free available chlorine) remaining in water, sewage, or industrial wastes at the end of a specified contact period following chlorination.

CLARIFICATION: Any process or combination of processes to reduce the concentration of suspended matter in a liquid.

CLARIFIERS: Settling tanks. The purpose of a clarifier is to remove settleable solids by gravity, or colloidal solids by coagulation.

COAGULATION: Process by which chemicals (coagulants) are added to an aqueous system, to render finely divided, dispersed matter with slow or negligible settling velocities into more rapidly settling aggregates. Forces that cause dispersed particles to repel each other are neutralized by the coagulants.

COD: Chemical Oxygen Demand

COLIFORM-GROUP BACTERIA: A group of bacteria predominantly inhabiting the intestines of man or animal, but also occasionally found elsewhere. Used as an indicator of human fecal contamination.

COLLOIDS: The finely divided suspended matter which will not settle, and the apparently dissolved matter which may be transformed into suspended matter by contact with solid surfaces or precipitated by chemical treatment. Substances which are soluble as judged by ordinary physical tests, but will not pass through a parchment membrane.

COMPOSTING: Controlled decomposition of organic matter under aerobic conditions by which material is transformed into humus. The process is normally exothermic resulting in a rise in temperature.

DENITRIFICATION: The reduction of nitrate to nitrogen gas by denitrifying organisms.

DETENTION TIME: Average period of time a fluid element is retained in a basin or tank before discharge.

DIALYSIS: Separation of a colloid from a substance in true solution, by allowing the solution to diffuse through a semi-permeable membrane.

DIGESTION: The controlled decomposition of organic substances, normally under anaerobic conditions.

DIGESTER: The unit in which anaerobic digestion takes place, and the unit often has the capability of retaining the biogas produced by anaerobic digestion.

DISINFECTION: Killing pathogenic microbes on or in a material without necessarily sterilizing it.

DISSOLVED OXYGEN (DO): The oxygen dissolved in water, wastewater, or other liquid, usually expressed in milligrams per liter (mg/L), parts per million (ppm), or percent of saturation.

DISSOLVED SOLIDS: Theoretically, the anhydrous residues of the dissolved constituents in water. Actually, the term is defined by the method used in determination.

DO: Dissolved Oxygen

DUAL MEDIA FILTRATION: Filtration process that uses a bed composed of two distinctly different granular substances (such as anthracite coal and sand), as opposed to conventional filtration through sand only.

ECOLOGY: The branch of biology that deals with the mutual relations of living organisms and their environments, and the relations of organisms to each other.

ECOSYSTEM: The functioning together of the biological community and the non-living environment.

ELECTRICAL CONDUCTIVITY: Reciprocal of the resistance in ohms measured between opposite faces of a centimeter cube of an aqueous solution at a specified temperature. Expressed as microhms per centimeter in degrees Celsius.

EFFLUENT: Sewage, water, or other liquid, partially or completely treated or in its natural state, flowing out of a reservoir, basin, or treatment plant.

ELECTROSTATIC PRECIPITATION: A process in which particles are collected by means of electric charge.

EMISSION: In environmental work, a reference to gaseous discharges to the atmosphere as opposed to effluent which refers to liquid and solid discharges.

EUTROPHIC WATERS: Waters with a good supply of nutrients; they may support rich organic production, such as algal blooms.

EUTROPHICATION: Process whereby lakes or streams become enriched with biological nutrients, usually nitrogen and phosphorus.

EXTENDED AERATION: A modification of the activated sludge process which provides for aerobic sludge digestion within the aeration system.

FATS: Triglyceride esters of fatty acids. Erroneously used as synonym for grease.

FECAL COLIFORM: An indicator organism for evaluating the microbiological suitability of the water.

FLOC: An agglomeration of finely divided or colloidal particles.

GREASE: In wastewater, a group of substances, including fats, waxes, fatty acids, calcium and magnesium soaps, mineral oils, and certain other nonfatty materials. The type of solvent and method used for extraction should be stated for quantification.

HARDNESS: Characteristic of water imparted by salts of calcium, magnesium, and iron (such as bicarbonates, carbonates, sulfates, chlorides, and nitrates), which causes curdling of soap, deposition of scale in boilers, damage in some industrial processes, and sometimes objectionable taste. It may be determined by a standard laboratory procedure or computed from the amounts of calcium, magnesium, iron, aluminum, manganese, barium, strontium, and zinc, and is expressed as equivalent calcium carbonate.

HYDRAULIC LOADING: Quantity of flow passing through a column or packed bed, expressed in the units of volume per unit time per unit area; e.g., $m^3/m^2 \cdot s$ ($gal/min/ft^2$)

IMMEDIATE OXYGEN DEMAND: Oxygen consumed by a wastewater sample within a brief period (1 to 2 minutes) after aeration commences.

INCINERATION: With reference to gaseous materials, an abatement technique where the streams are heated to a specified temperature for a significant length of time to enable combustion of the products.

INFLUENT: Water, wastewater, or other liquid flowing into a reservoir, basin, or treatment plant.

INTEREST: The cost of borrowing money. It is a function of the unrepaid principal and is expressed as a per cent per year.

IOD: Immediate Oxygen Demand

LIMNOLOGY: The study of the physical, chemical, and biological aspects of inland waters.

MIXED LIQUOR: Mixture of activated sludge and wastewater undergoing activated sludge treatment in the aeration tank.

MIXED LIQUOR SUSPENDED SOLIDS (MLSS): Concentration of suspended solids carried in the aeration basin of an activated sludge process.

MLSS: Mixed Liquor Suspended Solids.

MPN: Most probable number. Expressed as density of organisms per 100 mL.

NITRIFICATION: The bacterial oxidation of nitrogenous compounds, such as the production of nitrite and nitrate from ammonia and proteinaceous substances.

NONSETTLEABLE SOLIDS: Suspended matter that does not settle or float to the surface of water in a period of 1 hour.

NUTRIENT: Any substance assimilated by an organism which promotes growth and replacement of cellular constituents.

OIL AND GREASES: Oils and greases are determined by multiple solvent extractions of the filterable portion of a sample of waste water; therefore, floating oils and greases are not included in the analysis. Several solvents are commonly used and each gives a different result with the same sample. Standardized tests are recommended, but there is much disagreement as to what constitutes the best method. Solvents such as hexane, ether, Freon, and carbon tetrachloride are used, and it is important that the solvent be specified. Oil and grease exert an oxygen demand, cause unsightly conditions, and can interfere with anaerobic biological treatment systems.

OLIGOTROPHIC WATERS: Waters with a small supply of nutrients; hence, they support little organic production.

ORGANIC MATTER: Chemical substances of animal or vegetable origin of basically carbon structure, comprising compounds consisting of hydrocarbons and their derivatives.

ORGANIC NITROGEN: Nitrogen combined in organic molecules, such as protein, amines, and amino acids.

OVERFLOW RATE: One of the criteria for the design of settling tanks in treatment plants, expressed in cubic meters per day per square meter (gallons per day per square foot) of surface area in the settling tank.

OXIDATION: Addition of oxygen to a compound. More generally, any reaction involving the loss of electrons from an atom.

OXIDATION POND OR LAGOON: Basin used for retention of wastewater before final disposal, in which biological oxidation of organic material is effected by natural or artificially accelerated transfer of oxygen to the water from air.

OXIDIZING AGENTS: Any substance which can receive electrons and thereby cause some other chemical to increase in positive charge.

PARTICULATE MATTER: Any material except uncombined water which exists in a finely divided form as a liquid or solid.

pH: Unit used to describe acidity or alkalinity. A pH value of 7 is neutral; above 7 is alkaline and below 7 is acidic.

POPULATION EQUIVALENT: The total mass BOD in an industrial wastewater divided by the mass of BOD contributed per person per day to a domestic wastewater, i.e. 1000 kg of BOD in an industrial wastewater/0.114 kg per capita = 8,772 people.

PRIMARY TREATMENT: (a) First (sometimes only) major treatment in a wastewater treatment works, usually sedimentation; or (b) removal of a substantial amount of suspended matter, but little or no colloidal and dissolved matter.

PRIORITY POLLUTANT: One of 129 pollutants identified by the U.S. Environmental Protection Agency as being particularly toxic. This list of 129 pollutants includes 116 organic and 13 inorganic chemicals.

RECEIVING WATER: Surface waters which assimilate effluent discharge.

SANITARY LANDFILL: A controlled method of refuse disposal in which refuse is dumped on land in accordance to a preconceived plan, compacted and covered during and at the end of each day.

SANITARY SEWER: Sewer that carries liquid and water-carried human wastes from residences, commercial buildings, industrial plants, and institutions, together with minor quantities of storm, surface, and groundwater(s) that are not admitted intentionally. Significant quantities of industrial wastewater are not carried in sanitary sewers.

SCFM: Standard cubic feet per minute. Air flow corrected to predefined standard conditions of temperature and pressure, generally 32°F and one atmosphere in air pollution work.

SCRUBBER: In air pollution, a device in which a contaminated stream is contacted with a liquid to reduce contaminant emission.

SECONDARY WASTEWATER TREATMENT: Treatment of wastewater by biological methods after primary treatment by sedimentation.

SEDIMENTATION: Process of subsidence and deposition of suspended matter carried by water, wastewater, or other liquids, by gravity. Usually accomplished by reducing the velocity of the liquid to below the point at which it can transport the suspended material. Also called settling.

SEPTIC: Causing anaerobic biological activities due to insufficient oxygen present in wastewaters.

SELF-PURIFICATION: Natural processes occurring in a stream or other body of water resulting in the reduction of bacteria, satisfaction of the BOD, stabilization of organic constituents, replacement of depleted dissolved oxygen, and the return of the stream biota to normal. Also called natural purification.

SETTLEABLE SOLIDS: That matter in wastewater which will not stay in suspension during a preselected settling period, such as one hour, but either settles to the bottom or floats to the top.

SLUDGE: The slurry of settled particles resulting from the process of sedimentation.

SLUDGE VOLUME INDEX (SVI): Numerical expression of the settling characteristics of activated sludge. The ratio of the volume in milliliters of sludge settled from a 1,000-mL sample in 30 minutes to the concentration of mixed liquor in milligrams per liter multiplied by 1,000.

SOLIDS: Material in the solid state. *Total:* The solids in water, sewage, or other liquids; includes suspended and dissolved solids; all material remaining as residue after water has been evaporated. *Dissolved:* Solids present in solution. *Suspended:* Solids physically suspended in water, sewage, or other liquids. The quantity of material deposited when a quantity of water, sewage, or liquid is filtered through an asbestos mat or glass fiber filter. *Volatile:* The quantity of solids in water, sewage, or other liquid lost on ignition of total solids.

SOLIDS RETENTION TIME (SRT): The average residence time of suspended solids in a biological waste treatment system, equal to the total weight of suspended solids in the system divided by the total weight of suspended solids leaving the system per unit time (usually per day).

SS: Suspended solids.

STABILIZATION PONDS: Ponds or lagoons used in treatment of sewage, also called oxidation ponds or stabilization lagoons. These may be either anaerobic (due to high sewage loads and lack of oxygen), aerobic (with oxygen provided by algae), or more commonly facultative (being aerobic in the surface layers and anaerobic toward the bottom).

SUSPENDED SOLIDS (SS): Suspended solids are the suspended material that can be removed from wastewaters by laboratory filtration excluding coarse or floating solids that can be screened or settled out readily. Suspended solids are a vital and easily determined measure of pollution and also a measure of the material that may settle out in slow-moving streams. Both organic and inorganic materials are measured by the SS test.

SVI: Sludge Volume Index.

TOC: Total Organic Carbon.

TOTAL ORGANIC CARBON (TOC): Measure of the amount of organic material in a water sample, expressed in milligrams per liter of carbon. Measured by carbonaceous analyzer in which the organic compounds are catalytically oxidized to CO_2 and measured by an infrared detector. Frequently applied to wastewaters.

TOTAL SOLIDS: Sum of dissolved and undissolved constituents in water or wastewater, usually expressed in milligrams per liter.

TSS: Total suspended solids. Amount of solids separated by filtration of a sample of wastewater.

TURBIDITY: Condition in water or wastewater caused by the presence of suspended matter, resulting in the scattering and absorption of light rays. Measure of fine suspended matter in liquids. Analytical quantity, usually expressed in Jackson turbidity units (Jtu), determined by measurements of light diffraction.

ULTIMATE BIOCHEMICAL OXYGEN DEMAND (UBOD): Quantity of oxygen required to satisfy completely biochemical oxygen demands.

VAPOR PRESSURE: (a) The pressure exerted by a vapor in a confined space. It is a function of the temperature. (b) The partial pressure of water vapor in the atmosphere. (c) Partial pressure of any liquid.

VENTURI SCRUBBER: A scrubber in which gas velocity is increased in the presence of a liquid due to a decrease in cross sectioned area of the duct causing particulate matter to be captured by impaction into the liquid.

VOLATILE SOLIDS: Quantity of solids in water, wastewater, or other liquids lost on ignition of the dry solids at 600°C.

VSS: Volatile suspended solids.

APPENDIX B

AIR EMISSIONS FROM VARIOUS PETROCHEMICAL PROCESSING PLANTS

Table B-1. Air Emissions from the Acetaldehyde via Ethylene Process (kg/kg product produced) (Pervier et al. 1974 a, b, c, d)

Emission	Source		Total
	Regenerator Off-Gas Scrubber Vent	Light Ends Scrubber Vent	
Hydrocarbons	.00045	.00047	.00092
Particulates	0	0	0
NO_x	0	0	0
SO_x	0	0	0
CO	0	0	0

Table B-2. Air Emissions from the Acetaldehyde via Ethanol Process (kg/kg product produced) (Pervier et al. 1974 a, b, c, d)

Emission	Source		
	Absorbent Vent	Fuel for Reactor Start-Up	Ethanol Storage Vent
Hydrocarbons	0	Negligible	Negligible
Particulates	0	Negligible	Negligible
NO_x	0	Negligible	Negligible
SO_x	0	Negligible	Negligible
CO	.00276		

Table B-3. Air Emissions from the Acetic Acid via Methanol Process (kg/kg product produced) (Pervier et al. 1974 a, b, c, d)

Emission	Source		Total
	Purification Section	Acid Recovery	
Hydrocarbons			
Particulates & Aerosols			
NO_x	- See below -		0.00003
SO_x			
CO			

<u>Note</u>:

All waste gas streams leaving this process are flared or incinerated. No data on combusted gases are available; however, these gases contain no NO_x or SO_x or particulates, and if incinerated properly, should give only CO_2 and H_2O as off gas, except about 20 - 40 ppm of NO_x from atmospheric nitrogen, as indicated in the total column.

Table B-4. Air Emissions from the Acetic Acid via Butane Process (kg/kg product produced) (Pervier et al. 1974 a, b, c, d)

Emission	Source				Total
	Reactor Stripping Section	Liquid Waste Incineration	Venturi Scrubber and Purification Section Relief Valves	Tank Vents & Fugitive Emissions	
Hydrocarbons	.03824	0	Negligible	.00087	.03911
Particulates	0	0	0	0	0
NO_x	0	.00004	0	0	.00004
SO_x	0	0	0	0	0
CO	.01354	0	0	0	.01354

Table B-5. Air Emissions from the Acetic Acid via Acetaldehyde Oxidation Process (kg/kg product produced) (Pervier et al. 1974 a, b, c, d)

Emission	Source	Total
	Absorber Vent	
Hydrocarbons	.00983	.00983
Particulates	0	0
NO_x	0	0
SO_x	0	0
CO	.00202	.00202

Table B-6. Air Emissions from the Acetic Anhydride Process (kg/kg product produced) (Pervier et al. 1974 a, b, c, d)

Emission	Source			Total
	Absorber-Scrubber Tail Gas	Tar Incinerator Gas	Fugitive Emissions	
Hydrocarbons	0.002729	0		0.002729
Particulates & Aerosols	0	TR		TR
NO_x	0	0	None Given	0
SO_x	0	0		0
CO	0.004968	a		0.004968

[a]Compound present but no analysis available.

Table B-7. Air Emissions from the Adipic Acid Process (kg/kg product produced) (Pervier et al. 1974 a, b, c, d)

Emission	Source				Total
	Reactor Off-Gas	Product Purification and Nitric Acid Recovery	"Finished" Product Operations - Conveying, Drying, etc.	Fugitive Emissions	
Hydrocarbons				Negligible	0
Particulates		.00005	.00010	Negligible	.00015
NO_x	.0037	.017		Negligible	.0207
SO_x				Negligible	0
CO	.00010			Negligible	.00010

Table B-8. Air Emissions from the Adiponitrile via Butadiene Process (kg/kg product produced) (Pervier et al. 1974 a, b, c, d)

Emission	Source[a]					Total
	Chlorination	Cyanide Synthesis	Cyanation and Isomerization	Hydrogenation	Boiler House Combustion of Liquid Waste	
Hydrocarbons	.01791		.00773			.02564
Particulates & Aerosols			.00381	.00004	.00688	.01073
NO_x		.0379	.02120		.05686	.11596
SO_x						
CO						

[a]Emissions from individual sections include fugitive emissions.

Table B-9. Air Emissions from the Adiponitrile via Acid Process (kg/kg product produced) (Pervier et al. 1974 a, b, c, d)

Emission	Source			Total
	Ammonia Recovery Section	Product Fractionation Section	Fugitive Emissions	
Hydrocarbons				
Particulates & Aerosols		.00178		.00178
NO_x	.000137		Negligible	.000137
SO_x				
CO				

Table B-10. Air Emissions from the Polypropylene Process (kg/kg product produced) (Pervier et al. 1974 a, b, c, d)

Emission	Source					Total
	Reaction Area	HC Recycle	Recovery Area	Materials Handling	Fugitive Emissions	
Hydrocarbons	0.0003	0.0008	0.0053	0.0105	0.0150[a]	0.0319
Particulates	0.0001	None	None	None	None	0.0001
NO_x	None					0.00
SO_x						0.00
CO						0.00

[a]Assumed to be a true fugitive figure for a new, "clean", well-maintained installation.

Table B-11. Air Emissions from the Polystyrene Process (kg/kg product produced) (Pervier et al. 1974 a, b, c, d)

	Source						
Emission	Feed Preparation Section	Reactor Vent	Solvent Recovery Section	Conveying Operations	Fired Heater Flue Gas	Fugitive Emissions	Total
Hydrocarbons	.00065	.00334	.00184				.00583
Particulates				.00010			.00010
NO_x						Negligible	0
SO_x					.00033		.00033
CO							0

Table B-12. Air Emissions from the Polyvinyl Chloride Process (kg/kg product produced) (Pervier et al. 1974 a, b, c, d)

Emission	Source									Total
	Monomer Storage Area	Reaction Stripping and "Slurry"or "Blend Tank Areas	Centri- fuge Area	Vinyl Chloride Vent Condenser	Dryer Collec- tor Effluent	Vinyl Chloride from Vinyl Chloride Still Bottoms	Silo Collec- tor Fines	Bagger Collec- tor Fines	Bulk Loading Collec- tor Fines	Fugitive Emissions	
Hydrocarbons	0.0009			0.0010	0.0100	0.0002				0.0020	0.0141
Particulates					0.0020		0.0003	0.0001	0.0003		0.0027
NO_x											0.00
SO_x											0.00
CO											0.00

Table B-13. Air Emissions from the Styrene-Butadiene Rubber Process (kg/kg product produced) (Pervier et al. 1974 a, b, c, d)

	Source						
Emission	Monomer Recovery	Polymer Extrusion and Drying	Carbon Black Handling	Packaging	Fugitive Emissions	Heat and Power Gen.	Total
Hydrocarbon	.00010	.00100			.00100		.00210
Particulates & Aerosols		.00002	.00010	.00002	.00021		.00035
NO_x							0
SO_x						.00020	.00020
CO							0

Table B-14. Air Emissions from the Vinyl Acetate via Acetylene Process (kg/kg product produced) (Pervier et al. 1974 a, b, c, d)

Emission	Source		Total
	Light Ends Vent	Incineration of Liquid Waste	
Hydrocarbons	.02996	0	.02996
Particulates	0	Negligible	0
NO_x	0	Negligible	0
SO_x	0	0	0
CO	0	0	0

Table B-15. Air Emissions from the Vinyl Acetate via Adipic Acid Process (kg/kg product produced) (Pervier et al. 1974 a, b, c, d)

Emission	Source		Total
	Light Ends Flare	Heavy Ends Thermal Oxidizer	
Hydrocarbons	0	0	0
Particulates	0	0	0
NO_x	Trace	Trace	Trace
SO_x	0	0	0
CO	0	0	0

Table B-16. Air Emissions from the Vinyl Chloride via EDC Pyrolysis Process (kg/kg product produced) (Pervier et al. 1974 a, b, c, d)

Pollutant	Source							Total
	EDC Drier and Vaporizer plus VCM Drier	Cracking Furnace	Quench Tower	HCL Tower	"Light Ends" Reject	"Heavy Ends" Reject	Storage Losses and Fugitive Emissions	
Hydrocarbons	0.00100		0.00005	0.00010	0.00100	0.00100	0.00010	0.00325
Particulates			0.00001	0.00010				0.00011
NO_x		TR						TR
SO_x		TR						TR
CO								

Table B-17. Air Emissions from the Maleic Anhydride Process (kg/kg product produced) (Pervier et al. 1974 a, b, c, d)

Emission	Source	
	Scrubber Vent	Storage Losses and Fugitive Emissions
Hydrocarbons	.086	a
Particulates		b
NO_x		
SO_x		
CO	.670	

[a] There will be small amounts of hydrocarbon emissions from storage tanks but the amount is not available.

[b] Fugitive dust emissions will mostly be composed of maleic anhydride powder from the pelletizing, handling and storage operations of maleic anhydride. The amount is not indicated and will vary from plant to plant depending on operations.

Table B-18. Air Emissions from the Nylon 6 Process (kg/kg product produced) (Pervier et al. 1974 a, b, c, d)

Emission	Source					Total
	Mixing Tank Vents	Reactor Vent	Pellet Formation Washing & Drying Vents	Furnace Cleaning	Caprolactam Recovery	
Hydrocarbons	0	0	0	0	0	0
Aerosols & Particulates	.00012	.00034	.00172	0	0.001	0.00318
NO_x	0	0	0	Trace	0	Trace
SO_x	0	0	0	0	0	0
CO	0	0	0	0	0	0

Table B-19. Air Emissions from the Nylon 6, 6 Process (kg/kg product produced) (Pervier et al. 1974 a, b, c, d)

Emission	Source					Total
	Evaporation Section	Reactor Section	Flasher Section	Finishing	Fugitive	
Hydrocarbons	0	0	0		0	0
Particulates & Aerosols	0.000333	0.002100	0.001100	0.000044	0	0.003577
NO_x	0	0	0	0	0	0
SO_x	0	0	0	0	0	0
CO	0	0	0	0	0	0

Table B-20. Air Emissions from the Oxo Process (kg/kg product produced) (Pervier et al. 1974 a, b, c, d)

Emission	Source						Total
	Reforming Furnaces Vent Gas	Oxo Reactor System Off-Gas	Catalyst System Vent Gas	Compressor Engine Exhausts	Distillation Purification System Vents	Heavy Liquid Incinerator Stack Gas	
Hydrocarbons	0.000009	0.000010	0.000040		0.002970	TR	0.003029
Particulates & Aerosols		0.000003	TR			TR	0.000003
NO_x		0.000040					0.000040
SO_x							
CO	0.000009	0.000240		0.011000			0.011249

Table B-21. Air Emissions from the Phenol Process (kg/kg product produced) (Pervier et al. 1974 a, b, c, d)

Emission	Source				Total
	Oxidation Section	Concentration Cleavage Section	Distillation Section	Fugitive Emissions	
Hydrocarbons	.0038	.0021	.0038	.0006	.0103
Particulates & Aerosols	None	None	None	None	None
NO_x	None	None	None	None	None
SO_x	None	None	None	None	None
CO	None	None	None	None	None

Table B-22. Air Emissions from the High Density Polyethylene Process (kg/kg product produced) (Pervier et al. 1974 a, b, c, d)

Emission	Source							Total
	Catalyst Prep.	Reactor	Solvent Recovery	Polymer Stripping	Product Conveying	Fugitive	Flare	
Hydrocarbons			0.0020	0.0090	.0030	.0200		.0340
Particulates					.0010			.0010
NO_x	Negligible	Negligible					.0001	0
SO_x								0
CO								0

Table B-23. Air Emissions from the Low Density Polyethylene Process (kg/kg product produced) (Pervier et al. 1974 a, b, c, d)

Emission	Source					Total
	Compresser Purge	Reactor	Materials Handling	Gas-Separation Recovery Operation, Fugitive Emissions	Flare[a]	
Hydrocarbons	0.001	Negligible	0.005	0.010	TR	0.016
Particulates		Negligible	0.0003	None	None	0.0003
NO_x		Negligible	None	None	<.0001	<0.0001
SO_x		Negligible	None	None	None	0
CO		Negligible	None	None	None	0

[a]Flares, where used, are either intermittent, or else served other plant processes.

Table B-24 Air Emissions from the Carbon Disulfide Process (kg/kg product produced) (Pervier et al. 1974 a, b, c, d)

Emission	Source				Total
	Furnace Flue Gas	Sulfur Storage	Emergency Flaring	Fugitive Emissions	
Hydrocarbons[a]	.0009			.00007	.00016
Particulates	.0004	.00028			.00032
NO_x	.00014				.00014
SO_x	Negligible		.00478		.00478
CO	Negligible				0

[a]Includes sulfur containing compounds excluding sulfur dioxides.

Table B-25. Air Emissions from the Cyclohexanone Process (kg/kg product produced) (Pervier et al. 1974 a, b, c, d)

Emission	Source					Total
	Oxidation	Hydrolysis and Saponification	Distillation	Dehydrogenation	Fugitive Emissions	
Hydrocarbons	.037	Negligible	.001	.001	Negligible	.039
Particulates & Aerosol		Negligible			Negligible	
NO_x		Negligible			Negligible	
SO_x		Negligible			Negligible	
CO	.042	Negligible	.001		Negligible	.043

Table B-26. Air Emissions from the Dimethyl Terephthalate Process (kg/kg product produced) (Pervier et al. 1974 a, b, c, d)

Emission	Total
Hydrocarbons	.03162
Particulates	.00047
NO_x	.00003
SO_x	.00036
CO	.01836

Table B-27. Air Emissions from the Ethylene Process (kg/kg product produced) (Pervier et al. 1974 a, b, c, d)

Emission	Furnace Decoking	Quench Tower	Acid Gas Removal	Compressors	Distillation Train	Emissions	Total
Hydrocarbons		.00001	.00001	.00010	.00050	.00010	.00072
Particulates	.00001						.00001
NO_x					.00001		.00001
SO_x			.00300				.00300
CO		.00001					.00001

Table B-28. Air Emissions from the Ethylene Dichloride (Direct) Process (kg/kg product produced) (Pervier et al. 1974 a, b, c, d)

Emission	Source			Total
	Scrubber Vent	Storage Losses	Fugitive Emissions	
Hydrocarbons	.00416	.00060	.00071	.00547
Particulates & Aerosols	0	0	0	0
NO_x	0	0	0	0
SO_x	0	0	0	0
CO	0	0	0	0

Table B-29. Air Emissions from the Formaldehyde (Silver Catalyst) Process (kg/kg product produced) (Pervier et al. 1974 a, b, c, d)

Emission	Source			Total
	Absorber Vent	Fractionator Vent	Fugitive Emissions	
Hydrocarbons	.00385	.00015	Negligible	.0040
Particulates	0	0	Negligible	0
NO_x	0	0	Negligible	0
SO_x	0	0	Negligible	0
CO	.0180	0	Negligible	.0180

Table B-30. Air Emissions from the Glycerol (Allyl Chloride) Process (kg/kg product produced) (Pervier et al. 1974 a, b, c, d)

Emission	Total
Hydrocarbons	0.06579
Particulates	None
NO_x	None
SO_x	None
CO	None

Table B-31. Air Emissions from the Hydrogen Cyanide (Andrussow) Process (kg/kg product produced) (Pervier et al. 1974 a, b, c, d)

Emission	Source				
	Process Off-Gas to Air Preheaters	Process Start-Up and Emergency Reactor Effluent	Process Sample Pot Vents	Process Tank & Process Off-Gas Vents	Fugitive Emissions
Hydrocarbons	0	0	.0012[c]	0	0
Particulates & Aerosols	0	0	0	0	0
NO_x	.0007[a]	.0001[b]	0	.0002[d]	0
SO_x	0	0	0	0	0
CO	0	0	0	0	0

[a] NO_x estimate, deriving from burning of hydrogen cyanide in hydrogen-rich process off-gas used for air preheater fuel.

[b] NO_x estimate from flaring of hydrogen cyanide and ammonia in streams and calculated on annual tonnage basis from instantaneous emission figures.

[c] Actually hydrogen cyanide.

[d] NO_x estimate from flaring of hydrogen cyanide content of streams.

Table B-32. Air Emissions from the Isocyanates via Amine Phosgenation Process (kg/kg product produced) (Pervier et al. 1974 a, b, c, d)

Emission	Source					Total
	Phosgene Decomposer	Isocyanate Scrubber Vent	Residual Gas Scrubber Vent	Miscellaneous Vents	Plant Flare	
Hydrocarbons	0.001134	0.000032	0.000015	TR	0	0.001181
Particulates & Aerosols	0.000033	0.000014	TR	0.000383	0.000249	0.000679
NO_x	0	0	0	0	0	0
SO_x	0	0	0.000015	0	0	0.000015
CO	0.004298	0.001159	0.077237	0	0	0.082694

APPENDIX C

ALTERNATIVE HAZARDOUS WASTE TREATMENT PROCESSES

Table C-1. A List of Alternative Hazardous Waste Treatment Processes

Listed below are the physical, chemical and biological unit treatment processes utilized from the Treatment Study. The selection of several unit treatment processes in proper sequence forms an Alternative Treatment Process.

Physical (P)

Air Stripping (AS)
Carbon Adsorption (CA)
Centrifugation (CENT)
Distillation (DIS)
Evaporation (EVAP)
Filtration (FIL)
Flocculation (FLOC)
Flotation (FLO)
Ion Exchange (IE)
Resin Adsorption (RA)
Reverse Osmosis (RO)
Sedimentation (SED)
Solvent Extraction (SE)
Steam Distillation (SD)
Steam Stripping (SS
Ultrafiltration (UF)
Crushing and Grinding (C&G)

Chemical (C)

Calcination (CAL) or Incineration (INC)
Catalysis (CAT)
Chlorinolysis (CL)
Electrolysis (EL)
Hydrolysis (HY)
Neutralization (NEU)
Oxidation (OX) - Includes Chlorination
Ozonation (OZ)
Precipitation (PPT)
Reduction (RED) - Includes De-
chlorination and Dehydrochlorination

Biological (B)

Activated Sludge (ASL)
Aerated Lagoon (AL)
Anaerobic Digestion (AD)
Composting (COM)
Trickling Filter (TF)
Waste Stabilization Pond (WSP)

Process Categories

I	Process is not applicable in a useful way to wastes of interest to this program.
II	Process might work in 5-10 years, but needs research effort first.
III	Process appears useful for hazardous wastes, but needs development work.
IV	Process is developed but not commonly used for hazardous wastes.
V	Process will be common to most industrial waste processors.

Table C-2. Alternative Treatment Processes for some Petrochemical Manufacturing Processes (Process Research, Inc. 1977)

Product and Typical Plant Size	Wastestream Components	Waste Generation KKg*/yr	Unit Treatment Process** (1)	(2)	(3)	(4)	Benefits Derived	Total Treatment Cost*** US$/KKg Waste	US$/KKg Prod.
Perchloroethylene 39,000 KKg/yr	Hexachlorobutadiene Chlorobenzenes Chloroethanes Chlorobutadiene Tars	12,000	(DIS) P,IV	(DIS) P,V			90 Percent reduction in Waste Recovery of Hexachlorobutadien Detoxification	-378.	-116.
Nitrobenzene	Crude Nitrated Aromatics	50	(SD) P,IV	(HY) C,III	(CAT) C,III		10 Percent reduction in waste 80 Percent converted to salable product (Nitrobenzene)	1930.	4.83
Chloromethane 50,000 KKg/yr	Hexachlorobenzene Hexachlorobutadiene	300	(DIS) P,IV (5) (DIS) P, IV	(CL) C,III	(DIS) P,IV	(NEU) C,V	75 Percent reduction in waste volume. Salable product (Carbon Tetra-chloride)	646.	3.84
Epichlorohydrin 75,000 KKg/yr	Epichlorohydrin Dichlorohydrin Chloroethers Trichloropropane Tars	4,000	(SE) P,III	(EVAP) P,III	(DIS) P,IV		75 Percent reduction in waste volume Recovery of Epichlorohydrin	0.50	0.03
Toluene Diisocyanate 27,500 KKg/yr	Polyurethane Ferric Chloride Isocyanates Tars	558	(HY) C,IV	(DIS) P,IV	(NEU) C,IV	(AL) P,III	Detoxification of waste. Partial waste recovery (Toluene Diamine)	428.	3.69

*KKg = 1 Metric Ton (MT)

**See abbreviations in Table 5.3

***Includes credit for material recovery where applicable

A minus sign indicates a cost credit.

Table C-2. (continued)

Product and Typical Plant Size	Wastestream Components	Waste Generation KKg*/yr	Unit Treatment Process** (1)	(2)	(3)	(4)	Benefits Derived	Total Treatment Cost*** US$/KKg Waste	US$/KKg Prod.
Vinyl Chloride Monomer 136,000 KKg/yr	1,2 Dichloroethane 1,1,2 Trichloroethane 1,1,1,2 Tetrachloroethane Tars	1,400	(DIS) P,III	(RED) C,IV	(RED) C,IV		80 Percent reduction in waste volume. Recovery of >900MT of chlorinated Hydrocarbon	-0.84	-0.01
Methyl Methacrylate 55,000 KKg/hr	Hydroquinone Polymeric Residues	4,730	Not applicable					---	---
Acrylonitrile 80,000 KKg/yr	Acrylonitrile Higher Nitriles	160	Not applicable				Not applicable	---	---
Kaleic Anhydride 11,000 KKg/yr	Maleic Anhydride Fumeric Acid Chromogenic Compounds Tars	333	Not applicable				Not applicable	---	---

*KKg = 1 Metric Ton (MT)
**See Abbreviations in Table 5.3
***Includes credit for material recovery where applicable.
A minus sign indicates a cost credit.

Table C-2. (continued)

Product and Typical Plant Size	Wastestream Components	Waste Generation KKg*/yr	Unit Treatment Process** (1)	(2)	(3)	(4)	Benefits Derived	Total Treatment Cost*** US$/KKg Waste	US$/KKg Prod.
Lead Alkyls 60,000 KKg/yr	Lead	30,000	(FIL) P,V	(RED) C,IV	(FIL) P,V	(CAL) C,V	Recovery of lead oxide Detoxification Reduction in waste volume	-47.[(a)]	-24.[(a)]
			TRAIN NO. 1						
Zthanolamines 14,000 KKg/yr	Triethanolamine (TEA) Tars	1,120	(CENT) P,V	(DIS) P,IV			Recovery of 280MT of TEA Low energy input	188.	15.
			TRAIN NO. 3						
			(SED) P,V	(DIS) P,IV			Same as above	128.	10.
Furfural 35,000 KKg/yr	Sulfuric Acid Tars & Polymers	19,600	(SED) P,V	(HY) C,III	(COM) P,V		Recovery of 1000MT Sulfuric Acid. Eliminate Landfill. Waste volume reduced 5 percent	50.	29.
Furfural 35,000 KKg/yr	Pines & Particulates From Stripped Hulls	350	(DIS) P,IV	(HY) C,III	(COM) P,V		Recovery of 150MT/yr furfural	Combined with Furfural Stream	
Fluorocarbon 80,000 KKg/yr	Antimony Pentachloride Carbon Tetrachloride Trichlorofluoromethane Organics	18	TRAIN NO. 1 (RED) C,IV	(DIS)(b) P,IV			Catalyst recovery Low energy input	5560.	1.25
			TRAIN NO. 2						
			(DIS) P,IV		(c)		Detoxification Reduction in waste volume	470.	0.11

(a) Includes credit for Lead recovery
(b) Decholorination
(c) Distillation
*KKg = 1 Metric Ton (MT)
**See Abbreviations in Table 5.3
***Includes credit for material recovery where applicable.

Table C-2. (continued)

Product and Typical Plant Size	Wastestream Components	Waste Generation KKg*/yr	Unit Treatment Process** (1)	(2)	(3)	(4)	Benefits Derived	Total Treatment Cost*** US$/KKg Waste	US$/KKg Prod.
Chlorotoluene 15,000 KKg/yr	Benzylchloride Benzotrichloride	15	Not applicable				Not applicable	---	---
Chlorobenzene 32,000 KKg/yr	Polychlorinated Aromatic Resinous Material	1,400	Not applicable				Not applicable	---	---
Atranines 20,000 KKg/yr	Water Sodium Chloride Insoluble Residues Caustic Cyanuric Acid	224,600	(NEU) C,V	(OZ) C,III	(AL) B,II	(EVAP) P,V	Detoxification Boiler feedwater generated. Salable Product. (Salt)	4.60	51.
Trifluralin 10,000 KKg/yr	Spent Carbon Fluoroaromatics Intermediates and Solvents	1,150	(C&G) P,V (c) (COM) P,IV	(EE) P,III	(CENT) P,IV	(DIS) P,V	Recovery of 200MT/yr of Coloroform. Reduction in Waste Volume of 50 percent Moderate energy utilization.	400.	46.

(c) Distillation
*KKg = 1 Metric Ton (MT)
**See Abbreviations in Table 5.3
***Includes credit for material recovery where applicable.

Table C-2. (continued)

Product and Typical Plant Size	Wastestream Components	Waste Generation KKg*/yr	Unit Treatment Process** (1)	(2)	(3)	(4)	Benefits Derived	Total Treatment Cost*** US$/KKg Waste	US$/KKg Prod.
Malathion 14,000 KKg/yr	Filter Aid Toluene Insoluble Residues Dimethyl Dithiophos- phoric Acid	1,826	(HY) C,V (5) (AL) B,III	(SS) P,IV	(SED) P,V	(COM) P,IV	Recovery of 1 MT/day of Toluene, detoxification	93.	12.
Malathion 14,000 KKg/yr	Malathion Toluene Impurities Sodium Hydroxide	14,350(W) 350(D)	(SED) P,V	(RA) P,III	(DIS) P,V		Recovery of Toluene and Malathion, Total reuse of water	-0.38	-0.39
Parathion 20,000 KKg/yr	Diethylthiophosphoric Acid	2,300	(SED) P,V	(UF) P,II	(FIL) P,V	(COM) P,IV	Recovery of 2000 MT/yr of sulfur Reduction of waste volume.	73.	8.35
Explosives 93,000 KKg/yr	Activated Carbon Nitrobodies	330(W) 200(D)	(SE) P,III	(DIS) P,IV	(CAL) C,V		Cost Savings because of carbon regeneration	558. (W) 977. (D)	2.05
Explosives 30,000 KKg/yr	Redwater	15,000	(INC) - Tampella Process C,V				Recovery of Sellite Est. $780,000. 99.5 percent reduction of waste	213.	104.

*KKg = 1 Metric Ton (MT)
**See Abbreviations in Table 5.3
***Includes credit for material recovery where applicable.
(W) Wet Basis
(D) Dry Basis

Table C-2. (continued)

Product and Typical Plant Size	Wastestream Components	Waste Generation KKg*/yr	Unit Treatment Process** (1) (2) (3) (4)	Benefits Derived	Total Treatment Cost*** $/KKg Waste	Total Treatment Cost*** $/KKg Prod.
Explosives 125,000 KKg/yr	Waste Explosives	250	TRAIN NO. 1 (C&G) (OX) (ASL) (AD) P,V C,IV P,IV B,IV	COD reduced 80-97 percent. Recovery of energy. Total destruction of explos.	1580.	3.20
			TRAIN NO. 2 (C&G) (RED) (FIL) (EVAP) P,V C,IV P,V P,V (5) (CAL) C,V	Moderate energy required Desensitizing of explos.	1930. (d)	3.85 (d)

(d) Reduction
*KKg = 1 Metric Ton (MT)
**See Abbreviations in Table 5.3
***Includes credit for material recovery where applicable.
(W) Wet Basis
(D) Dry Basis

Table C-3. A Comparison of Hazardous Waste Treatment Alternative Costs for some Petrochemical Processes (Process Research, Inc. 1977)

			Alternative* Treatment Processes				Sanitary Landfill		Chemical Landfill		Incineration	
Product and Typical Plant Size	Wastestream Components	Waste Generation KKg/year	Cost US$/KKg Waste	Cost Impact US$/KKg Prod.	Prod. Selling Price US$/KKg	Impact on Prod. Selling Price %	Cost US$/KKg Waste	Cost Impact US$/KKg Prod.	Cost US$/KKg Waste	Cost Impact US$/KKg Prod.	Cost US$/KKg Waste	Cost Impact US$/KKg Prod.
Perchloroethylene 39,000 KKg/yr	Hexachlorobutadiene Chlorobenzenes Chloroethanes Chlorobutadiene Tars	12,000	-378.	-116.	390.	-29.8	10.	3.15	48.	16.	45.	14.
Nitrobenzene 20,000 KKg/yr	Crude Nitrated Aromatics	50	1930.	4.83	510.	0.95	98.	0.24	157.	0.39	N.A.	N.A.
Chloromethane 50,000 KKg/yr	Hexachlorobenzene Hexachlorobutadiene Tars	300	646.	3.88	320.	1.21	97.	0.58	128.	0.77	280.(1)	1.70(1)
Epichlorohydrin 75,000 KKg/yr	Epichlorohydrin Dichlorohydrin Chloroethers Trichloropropane Tars	4,000	0.50	0.03	882.	0.003	17.	0.92	55.	2.90	84.	4.50
Toluene Diisocyanate 27,500 KKg/yr	Polyurethane Ferric Chloride Isocyanates Tars	558	428.	8.60	1124.	0.77	97.	2.08	156.	3.34	231.	4.70

N.A. - Not applicable
(1) One Shift Per Day
* Includes credit for materials recovery where applicable
A minus sign indicates a cost credit

Table C-3. (continued)

			Alternative* Treatment Processes				Sanitary Landfill		Chemical Landfill		Incineration	
Product and Typical Plant Size	Wastestream Components	Waste Generation KKg/year	Cost US$/KKg Waste	Cost Impact US$/KKg Prod.	Prod. Selling Price US$/KKg	Impact on Prod. Selling Price %	Cost US$/KKg Waste	Cost Impact US$/KKg Prod.	Cost US$/KKg Waste	Cost Impact US$/KKg Prod.	Cost US$/KKg Waste	Cost Impact US$/KKg Prod.
Vinyl Chloride Monomer 136,000 KKg/yr	1,2 Dichloroethane 1,1,2 Trichloroethane 1,1,1,2 Tetrachloroethane Tars	1,400	-0.86	-0.01	300.	0.003	17.	0.25	67.	0.94	208.	2.10
Methyl Methacry-late 55,000 KKg/yr	Hydroquinone Polymeric Residues	4,730	---	---	840.	N.A.	17.	1.43	76.	6.55	30.	2.50
Acrylonitrile 80,000 KKg/yr	Acrylonitrile Higher Nitriles	160	---	---	590.	N.A.	98.	0.19	158.	0.31	350.(1)	1.10(1)
Maleic Anhydride 11,000 KKg/yr	Maleic Anhydride Fumaric Acid Chromogenic Compounds Tars	333	---	---	810.	N.A.	98.	2.95	166.	5.02	363.	12.
Lead Alkyls 60,000 KKg/yr	Lead	30,000	-47.	24.	1440.	-1.64	7.	3.50	61.	31.	N.A.	N.A.

N.A. - Not applicable
(1) One Shift Per Day
* Includes credit for materials recovery where applicable
A minus sign indicates a cost credit

Table C-3. (continued)

Product and Typical Plant Size	Wastestream Components	Waste Generation KKg/year	Alternative* Treatment Processes: Cost US$/KKg Waste	Alternative* Treatment Processes: Cost Impact US$/KKg Prod.	Alternative* Treatment Processes: Prod. Selling Price US$/KKg	Alternative* Treatment Processes: Impact on Prod. Selling Price $	Sanitary Landfill: Cost US$/KKg Waste	Sanitary Landfill: Cost Impact US$/KKg Prod.	Chemical Landfill: Cost US$/KKg Waste	Chemical Landfill: Cost Impact US$/KKg Prod.	Incineration: Cost US$/KKg Waste	Incineration: Cost Impact US$/KKg Prod.
Ethanolamines 14,000 KKg/yr	Triethanolamine Tars	1,120	188. (TRAIN I)	15.	790.	1.91	18.	1.40	77.	6.20	120.	9.60
			128. (TRAIN II)	10.		1.30						
Furfural 35,000 KKg/yr	Sulfuric Acid Tars and Polymers	19,600	50.	29.	1035.	2.77	8.	4.50	74.	43.	32.	18.
Furfural 35,000 KKg/yr	Fines & Particulates From Stripped Hulls	350	Included in Furfural Stream			---	---	---	---	---	143.	1.40
Fluorocarbon 80,000 KKg/yr	Antimony Pentachloride Carbon Tetrachloride Trichlorofluoromethane Organics	18	5560. (TRAIN I)	1.25	1080.	0.12	98.	0.02	117.	0.03	N.A.	N.A.
			470. (TRAIN II)	0.11		0.01						
Chlorotoluene 15,000 KKg/yr	Benzylchloride Benzotrichloride	15	---	---	660.	---	98.	0.10	156.	0.17	N.A.	N.A.
Chlorobenzene 32,000 KKg/yr	Polychlorinated Aromatic Resinous Material	1,400	---	---	570.	---	17.	0.77	70.	3.10	97.	4.20

* Includes credit for materials recovery where applicable
N.A. - Not applicable

Table C-3. (continued)

Product and Typical Plant Size	Wastestream Components	Waste Generation KKg/year	Alternative* Treatment Processes				Sanitary Landfill		Chemical Landfill		Incineration	
			Cost US$/KKg Waste	Cost Impact US$/KKg Prod.	Prod. Selling Price US$/KKg	Impact on Prod. Selling Price %	Cost US$/KKg Waste	Cost Impact US$/KKg Prod.	Cost US$/KKg Waste	Cost Impact US$/KKg Prod.	Cost US$/KKg Waste	Cost Impact US$/KKg Prod.
Atrazine 20,000 KKg/yr	Water Sodium Chloride Insoluble Residues Caustic Cyanuric Acid	224,600	4.60	51.	4295.	1.2	6.	71.	N.A.	N.A.	N.A.	N.A.
Trifluralin 10,000 KKg/yr	Spend Carbon Fluoro-aromatics Intermediates and Solvents	1,150	400.	46.	12290.	0.97	18.	2.04	326.	32.	123.	14.
Malathion 14,000 KKg/yr	Filter Aid Toluene Insoluble Residues Dimethyl Dithiophosphoric Acid	1,826	93.	12.	2090.	0.57	18.	1.80	326.	43.	91.	12.
Malathion 14,000 KKg/yr	Malathion Toluene Impurities Sodium Hydroxide	14,350(W) 350(D)	-0.38	-0.39	2090.	-0.019	18.	0.44	76.	1.90	30.(W)	31.
Parathion 20,000 KKg/yr	Diethylthiophosphoric Acid	2,300	73.	8.35	1918.	0.44	17.	2.	70.	5.	69.(2)	7.90(2)

(2) Two Shifts Per Day
(W) Wet
(D) Dry
N.A. - Not applicable
* Includes credit for materials recovery where applicable

Table C-3. (continued)

Product and Typical Plant Size	Wastestream Components	Waste Generation KKg/year	Alternative* Treatment Processes: Cost US$/KKg Waste	Cost Impact US$/KKg Prod.	Prod. Selling Price US$/KKg	Impact on Prod. Selling Price $	Sanitary Landfill: Cost US$/KKg Waste	Cost Impact US$/KKg Prod.	Chemical Landfill: Cost US$/KKg Waste	Cost Impact US$/KKg Prod.	Incineration: Cost US$/KKg Waste	Cost Impact US$/KKg Prod.
Explosives 93,000 KKg/yr	Activated Carbon Nitrobodies	350(W) 200(D)	558. 977.	2.05	N.A.	N.A.	N.A.	N.A.	N.A.	N.A.	873.(W)(2) 1530.(D)(2)	3.30(2)
Explosives 30,000 KKg/yr	Redwater Nitrobodies of DNT	15,000	213.	106.	N.A.	N.A.	N.A.	N.A.	N.A.	N.A.	268.	134.
Explosives 125,000 KKg/yr	Waste Explosives	250	1580. (TRAIN I) 1930. (TRAIN II)	3.20 3.85	N.A.	N.A.	N.A.	N.A.	N.A.	N.A.	1105.	2.20

(2) Two Shifts Per Day
(W) Wet
(D) Dry
N.A. - Not applicable
* Includes credit for materials recovery where applicable

APPENDIX D

CONVERSION FACTORS

Table D-1. Conversion Factors.

INSTRUCTIONS ON USE: TO CONVERT, MULTIPLY IN DIRECTION SHOWN BY ARROWS

	SI UNITS	-->	<--	U.S. UNITS
Length	centimeter	0.3937	2.5400	inch
	centimeter	0.032808	30.480	foot
	meter	39.3701	2.540×10^{-2}	inch
	meter	3.2808	0.30480	foot
	meter	1.0936	0.91441	yard
	kilometer	3,280.833	3.0480×10^{-4}	foot
	kilometer	0.6214	1.6093	mile
Area	$centimeter^2$	0.1550	6.4516	$inch^2$
	$meter^2$	10.7639	9.2903×10^{-2}	$foot^2$
	$meter^2$	1.1960	0.83612	$yard^2$
	$meter^2$	2.4711×10^{-4}	4046.78	acre
	$meter^2$	3.8610×10^{-7}	2.5900×10^{6}	$mile^2$
	$kilometer^2$	1.0764×10^{7}	9.29023×10^{-8}	$foot^2$
	$kilometer^2$	247.1044	4.0469×10^{-3}	acre
	$kilometer^2$	0.3861006	2.59000	$mile^2$
	hectare	107,638.7	9.290339×10^{-6}	$foot^2$
	hectare	2.47104	0.40468	acre

Table D-1. (continued)

INSTRUCTIONS ON USE: TO CONVERT, MULTIPLY IN DIRECTION SHOWN BY ARROWS

	SI UNITS	-->	<--	U.S. UNITS
Volume	$centimeter^3$	0.06102	16.3934	$inch^3$
	$centimeter^3$	3.5314×10^{-5}	2.8317×10^{4}	$foot^3$
	$centimeter^3$	2.6417×10^{-4}	3.7854×10^{3}	gallon
	$meter^3$	61,023.38	1.638716×10^{-5}	$inch^3$
	$meter^3$	35.3147	2.83168×10^{-2}	$foot^3$
	$meter^3$	1.3079	0.76458	$yard^3$
	$meter^3$	264.1720	3.7854×10^{-3}	gallon
	$meter^3$	8.3865	0.11924	barrel
	$meter^3$	8.1071×10^{-4}	1,233.487	acre-foot
	liter	33.8143	0.0295733	ounce
	liter	1.05668	0.946360	quart
	liter	0.2642	3.7853	gallon
	liter	61.025	0.016387	$inch^3$
	liter	0.0353	28.329	$foot^3$
Mass	milligram	0.015432	64.8004	grain
	milligram	3.5274×10^{-5}	28,349.49	ounce
	milligram	2.2046×10^{-6}	4.536×10^{5}	pound
	gram	0.035274	28.34949	ounce
	gram	0.002205	453.6	pound
	kilogram	2.2046	0.4536	pound
	kilogram	0.0011023	907.194	ton

Table D-1. (continued)

INSTRUCTIONS ON USE: TO CONVERT, MULTIPLY IN DIRECTION SHOWN BY ARROWS

	SI UNITS	-->	<--	U.S. UNITS
Velocity	meters/second	3.2808	0.304804	feet/second
	kilometers/sec	2.2369	0.44705	miles/hour
Acceleration	meters/second^2	3.2808	0.30480	feet/second^2
	meters/second^2	39.3701	2.5400×10^{-2}	inches/second^2
Temperature	Celsius (^oC)	$1.8(^oC) + 32$	$\frac{(^oF) - 32}{1.8}$	Fahrenheit (^oF)
	Kelvin (^oK)	$1.8(^oK) - 459.67$	$\frac{(^oF) + 459.67}{1.8}$	Fahrenheit (^oF)
Flow Rate	liters/second	15.8508	0.063088	gallons/minute
	liters/second	22,824.5	4.38126×10^{-5}	gallons/day
	liters/second	0.0228245	43.8126	million gallons/day
	liters/second	0.035316	28.3158	feet^3/second
	meters^3/second	15,850.3	6.3088×10^{-5}	gallons/minute
	meters^3/second	2.28245×10^7	4.38126×10^{-8}	gallons/day
	meters^3/second	22.8245	4.38126×10^{-2}	million gallons/day
	meters^3/second	35.316	0.028316	feet^3/second
Energy	joule	0.9478	1.0551	British thermal unit
	joule	2.778×10^{-7}	3.600×10^6	kilowatt-hour
	joule	0.7376	1.3557	foot-pound (force)
	joule	1.000	1.0000	watt-second
	joule	0.2388	4.1876	calorie
	joule	2.778×10^{-4}	3,599.71	watt-hour

Table D-1. (continued)

INSTRUCTIONS ON USE: TO CONVERT, MULTIPLY IN DIRECTION SHOWN BY ARROWS

	SI UNITS	-->	<--	U.S. UNITS
Power	watt	0.7376	1.35575	foot-pounds(force)/second
	watt	0.001341	745.7	horsepower
	watt	9.478×10^{-4}	1,055.1	British thermal units/second
	watt	0.014333	69.7691	calories/minute
Pressure	pascal	1.4504×10^{-4}	6,894.65	pounds(force)/inch^2
	pascal	2.0885×10^{-2}	47.88125	pounds(force)/foot^2
	pascal	2.9613×10^{-4}	3,376.895	inches of mercury (60°F)
	pascal	4.0187×10^{-3}	248.8367	inches of water (60°F)
	pascal	9.8687×10^{-6}	101,330	atmosphere

INDEX